D0207754

The IMA Volumes in Mathematics and its Applications

Volume 43

Series Editors
Avner Friedman   Willard Miller, Jr.

Institute for Mathematics and its Applications
IMA

The **Institute for Mathematics and its Applications** was established by a grant from the National Science Foundation to the University of Minnesota in 1982. The IMA seeks to encourage the development and study of fresh mathematical concepts and questions of concern to the other sciences by bringing together mathematicians and scientists from diverse fields in an atmosphere that will stimulate discussion and collaboration.

The IMA Volumes are intended to involve the broader scientific community in this process.

Avner Friedman, Director
Willard Miller, Jr., Associate Director

* * * * * * * * * *

IMA ANNUAL PROGRAMS

1982–1983 **Statistical and Continuum Approaches to Phase Transition**
1983–1984 **Mathematical Models for the Economics of Decentralized Resource Allocation**
1984–1985 **Continuum Physics and Partial Differential Equations**
1985–1986 **Stochastic Differential Equations and Their Applications**
1986–1987 **Scientific Computation**
1987–1988 **Applied Combinatorics**
1988–1989 **Nonlinear Waves**
1989–1990 **Dynamical Systems and Their Applications**
1990–1991 **Phase Transitions and Free Boundaries**
1991–1992 **Applied Linear Algebra**
1992–1993 **Control Theory and its Applications**
1993–1994 **Emerging Applications of Probability**

IMA SUMMER PROGRAMS

1987 **Robotics**
1988 **Signal Processing**
1989 **Robustness, Diagnostics, Computing and Graphics in Statistics**
1990 **Radar and Sonar**
1990 **Time Series**
1991 **Semiconductors**
1992 **Environmental Studies: Mathematical, Computational, and Statistical Analysis**

* * * * * * * * * *

SPRINGER LECTURE NOTES FROM THE IMA:

The Mathematics and Physics of Disordered Media
Editors: Barry Hughes and Barry Ninham
(Lecture Notes in Math., Volume 1035, 1983)

Orienting Polymers
Editor: J.L. Ericksen
(Lecture Notes in Math., Volume 1063, 1984)

New Perspectives in Thermodynamics
Editor: James Serrin
(Springer-Verlag, 1986)

Models of Economic Dynamics
Editor: Hugo Sonnenschein
(Lecture Notes in Econ., Volume 264, 1986)

Morton E. Gurtin   Geoffrey B. McFadden
Editors

On the Evolution of Phase Boundaries

With 17 Illustrations

Springer-Verlag
New York Berlin Heidelberg London Paris
Tokyo Hong Kong Barcelona Budapest

Morton E. Gurtin
Department of Mathematics
Carnegie-Mellon University
Pittsburgh, PA 15213
USA

Geoffrey B. McFadden
NIST
Gaithersburg, MD 20899
USA

Mathematics Subject Classifications (1991): 35, 73, 76, 80, 82

On the evolution of phase boundaries / Morton E. Gurtin, Geoffrey B. McFadden, editors.
p. cm. – (The IMA volumes in mathematics and its applications ; v. 43)
"Based on the proceedings of a workshop which was an integral part of the 1990–91 IMA program" – Foreword.
Includes bibliographical references.
ISBN 0-387-97803-8 (alk. paper). – ISBN 3-540-97803-8 (alk. paper)
1. Phase transitions (Statistical physics) – Congresses. 2. Boundary value problems – Congresses. 3. Differential equations, Partial – Congresses. 4. Mathematical physics – Asymptotic theory – Congresses. I. Gurtin, Morton E. II. McFadden, Geoffrey B. III. Series.
QC175.16.P505 1992
530.4'14 – dc20 92-125

Printed on acid-free paper.

© 1992 Springer-Verlag New York, Inc.
All rights reserved. This work may not be translated or copied in whole or in part without the written permission of the publisher (Springer-Verlag New York, Inc., 175 Fifth Avenue, New York, NY 10010, USA), except for brief excerpts in connection with reviews or scholarly analysis. Use in connection with any form of information storage and retrieval, electronic adaptation, computer software, or by similar or dissimilar methodology now known or hereafter developed is forbidden.
The use of general descriptive names, trade names, trademarks, etc., in this publication, even if the former are not especially identified, is not to be taken as a sign that such names, as understood by the Trade Marks and Merchandise Marks Act, may accordingly be used freely by anyone.
Permission to photocopy for internal or personal use, or the internal or personal use of specific clients, is granted by Springer-Verlag New York, Inc., for libraries registered with the Copyright Clearance Center (CCC), provided that the base fee of $0.00 per copy, plus $0.20 per page is paid directly to CCC, 21 Congress St., Salem, MA 01970, USA. Special requests should be addressed directly to Springer-Verlag New York, 175 Fifth Avenue, New York, NY 10010, USA.
ISBN 0-387-97803-8/1992 $0.00 + $0.20

Production managed by Hal Henglein; manufacturing supervised by Robert Paella.
Camera-ready copy provided by the IMA.
Printed and bound by Edwards Brothers, Inc., Ann Arbor, MI.
Printed in the United States of America.

9 8 7 6 5 4 3 2 1

ISBN 0-387-97803-8 Springer-Verlag New York Berlin Heidelberg
ISBN 3-540-97803-8 Springer-Verlag Berlin Heidelberg New York

530.414
On1
cop. 2

Math

The IMA Volumes
in Mathematics and its Applications

Current Volumes:

Volume 1: Homogenization and Effective Moduli of Materials and Media
Editors: Jerry Ericksen, David Kinderlehrer, Robert Kohn, J.-L. Lions

Volume 2: Oscillation Theory, Computation, and Methods of Compensated Compactness
Editors: Constantine Dafermos, Jerry Ericksen,
David Kinderlehrer, Marshall Slemrod

Volume 3: Metastability and Incompletely Posed Problems
Editors: Stuart Antman, Jerry Ericksen, David Kinderlehrer, Ingo Muller

Volume 4: Dynamical Problems in Continuum Physics
Editors: Jerry Bona, Constantine Dafermos, Jerry Ericksen,
David Kinderlehrer

Volume 5: Theory and Applications of Liquid Crystals
Editors: Jerry Ericksen and David Kinderlehrer

Volume 6: Amorphous Polymers and Non-Newtonian Fluids
Editors: Constantine Dafermos, Jerry Ericksen, David Kinderlehrer

Volume 7: Random Media
Editor: George Papanicolaou

Volume 8: Percolation Theory and Ergodic Theory of Infinite Particle Systems
Editor: Harry Kesten

Volume 9: Hydrodynamic Behavior and Interacting Particle Systems
Editor: George Papanicolaou

Volume 10: Stochastic Differential Systems, Stochastic Control Theory and Applications
Editors: Wendell Fleming and Pierre-Louis Lions

Volume 11: Numerical Simulation in Oil Recovery
Editor: Mary Fanett Wheeler

Volume 12: Computational Fluid Dynamics and Reacting Gas Flows
Editors: Bjorn Engquist, M. Luskin, Andrew Majda

Volume 13: Numerical Algorithms for Parallel Computer Architectures
Editor: Martin H. Schultz

Volume 14: Mathematical Aspects of Scientific Software
Editor: J.R. Rice

Volume 15: Mathematical Frontiers in Computational Chemical Physics
Editor: D. Truhlar

Volume 16: Mathematics in Industrial Problems
by Avner Friedman

Volume 17: Applications of Combinatorics and Graph Theory to the Biological and Social Sciences
Editor: Fred Roberts

Volume 18: q-Series and Partitions
Editor: Dennis Stanton

Volume 19: Invariant Theory and Tableaux
Editor: Dennis Stanton

Volume 20: Coding Theory and Design Theory Part I: Coding Theory
Editor: Dijen Ray-Chaudhuri

Volume 21: Coding Theory and Design Theory Part II: Design Theory
Editor: Dijen Ray-Chaudhuri

Volume 22: Signal Processing: Part I - Signal Processing Theory
Editors: L. Auslander, F.A. Grünbaum, J.W. Helton, T. Kailath, P. Khargonekar and S. Mitter

Volume 23: Signal Processing: Part II - Control Theory and Applications of Signal Processing
Editors: L. Auslander, F.A. Grünbaum, J.W. Helton, T. Kailath, P. Khargonekar and S. Mitter

Volume 24: Mathematics in Industrial Problems, Part 2
by Avner Friedman

Volume 25: Solitons in Physics, Mathematics, and Nonlinear Optics
Editors: Peter J. Olver and David H. Sattinger

Volume 26: Two Phase Flows and Waves
Editors: Daniel D. Joseph and David G. Schaeffer

Volume 27: Nonlinear Evolution Equations that Change Type
Editors: Barbara Lee Keyfitz and Michael Shearer

Volume 28: Computer Aided Proofs in Analysis
Editors: Kenneth Meyer and Dieter Schmidt

Volume 29: Multidimensional hyperbolic problems and computations
Editors Andrew Majda and Jim Glimm

Volume 30: Microlocal Analysis and Nonlinear Waves
Editors: Michael Beals, R. Melrose and J. Rauch

Volume 31: Mathematics in Industrial Problems, Part 3
by Avner Friedman

Volume 32: Radar and Sonar, Part 1
by Richard Blahut, Willard Miller, Jr. and Calvin Wilcox

Volume 33: Directions in Robust Statistics and Diagnostics: Part 1
Editors: Werner A. Stahel and Sanford Weisberg

Volume 34: Directions in Robust Statistics and Diagnostics: Part 2
Editors: Werner A. Stahel and Sanford Weisberg

Volume 35: Dynamical Issues in Combustion Theory
Editors: P. Fife, A. Liñán and F.A. Williams

Volume 36: Computing and Graphics in Statistics
Editors: Andreas Buja and Paul Tukey

Volume 37: Patterns and Dynamics in Reactive Media
Editors: Harry Swinney, Gus Aris and Don Aronson

Volume 38: Mathematics in Industrial Problems, Part 4
by Avner Friedman

Volume 39: Radar and Sonar, Part 2
Editors: F. Alberto Grünbaum, Marvin Bernfeld and Richard E. Blahut

Volume 40: Nonlinear Phenomena in Atmospheric and Oceanic Sciences
Editors: G.F. Carnevale and R.T. Pierrehumbert

Volume 41: Chaotic Processes in the Geological Sciences
Editor: David Yuen

Volume 42: Partial Differential Equations with Minimal Smoothness and Applications
Editors: B. Dahlberg, E. Fabes, R. Fefferman, D. Jerison, C. Kenig and J. Pipher

Volume 43: On the Evolution of Phase Boundaries
Editors: M.E. Gurtin and G. McFadden

Forthcoming Volumes:

1989-1990: *Dynamical Systems and Their Applications*

Twist Mappings and Their Applications

Dynamical Theories of Turbulence in Fluid Flows

Summer Program 1990: *Time Series in Time Series Analysis*

Time Series (2 volumes)

1990-1991: *Phase Transitions and Free Boundaries*

Shock Induced Transitions and Phase Structures

Microstructure and Phase Transitions

Statistical Thermodynamics and Differential Geometry of Microstructured Material

Free Boundaries in Viscous Flows

Variational Problems

Degenerate Diffusions

Summer Program 1991: *Semiconductors*

Semiconductors (2 volumes)

1991-1992: *Phase Transitions and Free Boundaries*

Sparse Matrix Computations: Graph Theory Issues and Algorithms

Combinatorial and Graph-Theoretic Problems in Linear Algebra

FOREWORD

This IMA Volume in Mathematics and its Applications

ON THE EVOLUTION OF PHASE BOUNDARIES

is based on the proceedings of a workshop which was an integral part of the 1990-91 IMA program on "Phase Transitions and Free Boundaries". The purpose of the workshop was to bring together mathematicians and other scientists working on the Stefan problem and related theories for modeling physical phenomena that occurs in two phase systems.

We thank M.E. Gurtin and G. McFadden for editing the proceedings.

We also take this opportunity to thank the National Science Foundation, whose financial support made the workshop possible.

Avner Friedman

Willard Miller, Jr.

PREFACE

A primary goal of the IMA workshop on the Evolution of Phase Boundaries from September 17–21, 1990 was to emphasize the interdisciplinary nature of contemporary research in this field, research which combines ideas from nonlinear partial differential equations, asymptotic analysis, numerical computation, and experimental science. The workshop brought together researchers from several disciplines, including mathematics, physics, and both experimental and theoretical materials science. Several of the formal talks involved extensions of the classical Stefan problem to include such effects as fluid flow, elasticity, and surface energy effects. In addition, recent work on alternate models was also highlighted. Much progress has been made on the mathematically-rigorous treatment of local models of interface motion based on dynamics determined solely by the geometry of the interface. Another promising area involves the formulation of phase-field models of phase transitions wherein the interfaces that separate the various phases are diffuse rather than sharp. Such approaches offer new settings for the development of practical numerical procedures for the solution of phase transition problems.

The ten papers in this volume span a wide cross-section of this research. Topics covered include the treatment of scaling laws that describe the coarsening or ripening behavior observed during the later stages of phase transitions; novel numerical methods for treating interface dynamics; the mathematical description of geometric models of interface dynamics; determination of the governing equations and interfacial boundary conditions in the context of fluid flow and elasticity.

The organizers would like to thank the staff of the IMA for their help with the details of the meeting and also for the preparation of this volume.

Morton E. Gurtin

Geoffrey B. McFadden

CONTENTS

PHASE FIELD EQUATIONS IN THE SINGULAR LIMIT OF SHARP INTERFACE PROBLEMS

GUNDUZ CAGINALP† AND XINFU CHEN‡

Abstract. In one of the singular limits as interface thickness approaches zero, solutions to the phase field equations formally approach those of a sharp interface model which incorporates surface tension. Here, we use a modification of the original phase field equations and prove this convergence rigorously in the one–dimensional and radially symmetric cases. Convergence to motion by mean curvature in another distinguished limit is also proved.

Key words. Phase field equations, Stefan problems, sharp interface, undercooling, surface tension.

1. Introduction

In this paper we consider solutions to a phase field model and prove that they are governed by solutions to a sharp interface model [see 7 and the references therein] (encompassing surface tension and kinetic undercooling) in a singular limit of vanishing interface thickness. The proof is restricted to the one–dimensional and radially symmetric cases, although a formal analysis indicates a more general result [6, 7].

The convergence of solutions to the phase field equations to those of sharp interface problems such as the Stefan model or the Hele–Shaw model was suggested by the asymptotic analysis [6–8]. It has already been proven rigorously in the special cases of the steady state problem [1, 9, 21] and the traveling wave problem [11]. Related theorems are also included in [4, 10, 17, 22].

The relevant sharp interface problems may be described as a material in a region $\Omega \subset \mathbb{R}^N$ which may be in either of two phases, e.g. solid and liquid (denoted by $-$ and $+$ respectively). The heat diffusion equation applies to each phase. Across the interface, Γ, the latent heat of fusion must be dissipated or absorbed in accordance with the conservation of energy. Since there is considerable practical, as well as theoretical, interest in these equations, we write these equations in the dimensional form as

$$\rho c_{\mathrm{spm}} \frac{\partial T}{\partial t} = K_{\mathrm{tc}} \Delta T \qquad \text{in } \Omega \setminus \Gamma, \tag{1.1}$$

$$\rho l_m v = K_{\mathrm{tc}} \big[\nabla T \cdot n \big]_+^- \qquad \text{on } \Gamma \tag{1.2}$$

where T is the (absolute) temperature, ρ the density, c_{spm} the specific heat per mass, [units of Energy(Mass $\cdot$ Degree)$^{-1}$], K_{tc} the thermal conductivity, [units of Energy(Area $\cdot$ Time $\cdot$ Temp Grad)$^{-1}$ = Energy(Length^{N-2} $\cdot$ Degree $\cdot$ Time)$^{-1}$], l_m

†Department of Mathematics and Statistics, University of Pittsburgh, Pittsburgh, PA 15260. The first author of this work is supported by the NSF Grant DMS-9002242.

‡School of Mathematics, University of Minnesota, Minneapolis, MN 55455. The second author would like to thank Professor Avner Friedman for his valuable discussions and the Sloan Doctoral Dissertation Fellowship (1990–1991) for its support.

the latent heat per mass, [units of Energy(Mass)$^{-1}$], v the normal velocity of the interface (positive if directed toward the liquid), n the unit vector normal to the interface (pointing to the liquid), and $[\ ]_{+}^{-}$ denotes the jump between solid and liquid. An additional condition which must be satisfied at the interface is given by

$$(1.3) \qquad [s]_E \left(T(x,t) - T_m\right) = -\sigma\kappa(x,t) - \alpha\sigma v \qquad \text{on } \Gamma$$

where s is the entropy per unit volume, [units of Energy(Length$^N \cdot$ Degree)$^{-1}$], $[s]_E$ is the difference in entropy (in equilibrium) per unit volume between the "+" phase and the "−" phase, κ the sum of principle curvatures at the point on Γ, σ is the surface tension, [units of Energy(Area)$^{-1}$ = Energy(Length^{N-1})$^{-1}$], and α is the relaxation scaling, [units of Time $\cdot$ Length^{-2}].

If σ is set to be zero in (1.3), then (1.1)–(1.3) is known as the classical Stefan model [27], a key feature of which is the distinguishability of phases based on the temperature alone. That is, $T(x,t) > T_m$ implies that the point belongs to liquid, and vice versa. This simple criterion for determining phases is no longer possible with the introduction of the more realistic physics embodied by (1.3) (with $\sigma \neq 0$) which allows for the possibility of supercooling (i.e., the presence of liquid below the melting temperature) and analogously superheating. The condition (1.3) with $\sigma \neq 0$ clearly is a stablizing influence on the shape of interface since surface tension multiplies the curvature term, thereby inhibiting interfaces with high curvature.

A convenient version of (1.1)–(1.3) which is often implemented in the physics literature uses a rescaled dimensionless temperature, u, diffusivity, D, and capillary length, d_0, defined by

$$u := \frac{T - T_m}{l_m / c_{\mathrm{spm}}}, \qquad D := \frac{K_{\mathrm{tc}}}{\rho c_{\mathrm{spm}}}, \qquad d_0 := \frac{\sigma}{[s]_E l_m / c_{\mathrm{spm}}}$$

(provided s is measured in the original, rather than dimensionless, temperature), so that the equations can be written in the form

$$(1.1') \qquad u_t = D\Delta u \qquad \text{in } \Omega \setminus \Gamma,$$

$$(1.2') \qquad v = D[\nabla u \cdot n]_{+}^{-} \qquad \text{on } \Gamma,$$

$$(1.3') \qquad u = -d_0 \kappa - \alpha d_0 v \qquad \text{on } \Gamma.$$

Equations (1.1′)–(1.3′) can be studied subject to appropriate initial and boundary conditions, e.g.,

$$(1.4) \qquad u(x,0) = u_0(x), \qquad x \in \Omega,$$

$$(1.5) \qquad u(x,t) = u_\partial(x), \qquad x \in \partial\Omega,\ t > 0.$$

Local existence and uniqueness of solutions to (1.1')–(1.3'), (1.4), (1.5) were recently proven by Chen and Reitich in [15].

It has been useful, from both theoretical and practical perspectives, to study (sharp interface) free boundary problems as a limit of problems with finite interfacial thickness which incorporate some of the physical structure of the interface. Toward this end, we consider a phase field system based on the free energy

$$\mathcal{F} := \int d^N x \left\{ \frac{1}{2}\xi^2 |\nabla\varphi|^2 - \frac{1}{a} G(\varphi) - 2u\gamma f(\varphi) \right\} \equiv \int F(u,\varphi) d^N x \tag{1.6}$$

which differs from the usual phase field equations with the insertion of a function $\gamma f(\varphi)$ replacing φ (see [8, p. 211]). Here the function $G(\varphi)$ is a symmetric double well potential with minima at ± 1, e.g., $\frac{1}{8}(\varphi^2 - 1)^2$, and $f(\varphi)$ is a function having the property that $f'(\pm 1) = 0$, thereby ensuring that the roots of

$$\frac{1}{a} G'(\varphi) + 2u\gamma f'(\varphi) = 0 \tag{1.7}$$

remain at ± 1. The variable φ is dimensionless and is called the phase or order parameter. Since the term $u\gamma f(\varphi)$ must have units of energy per volume with u and $f(\varphi)$ dimensionless, γ must have dimensions of energy per volume. The numerical value of γ will be related to the macroscopic parameters introduced in §2 below.

The Euler–Lagrange equations coupled with the nonequilibrium ansatz [8, p. 211]

$$\alpha\xi^2\varphi_t = -\delta\mathcal{F}/\delta\varphi$$

with $\alpha\xi^2$ as a relaxation time then implies the phase equation

$$\alpha\xi^2\varphi_t = \xi^2\Delta\varphi + \frac{1}{a} g(\varphi) + 2u\gamma f'(\varphi) \tag{1.8}$$

where $g(\varphi) := G'(\varphi)$. This equation is coupled with the conservation of energy equation

$$u_t + \frac{1}{2}\varphi_t = D\Delta u \tag{1.9}$$

using the units of (1.1′)–(1.3′), so that (1.8), (1.9) can be studied subject to suitable initial and boundary conditions. Although a variety of such conditions may be imposed, a key feature must include the vanishing of φ_t far from the interface so that the usual heat equation is attained in these regions.

The interface is now specified implicitly as the set of points on which φ vanishes; i.e.,

$$\Gamma(t) := \{x \in \Omega \mid \varphi(t,x) = 0\} \tag{1.10}$$

comprises the interface.

Within an appropriate scaling regime, we prove convergence (see Theorem 4.1) of solutions to the phase field equations (1.8), (1.9) to the sharp interface problem [(1.1)-(1.3)].

Equation (1.3′) alone with $u \equiv 0$ is known as the motion by mean curvature equation and is a formal limit of a phase equation such as (1.8) (see [2]). We prove convergence of (1.8), (1.9) to this problem within a different scaling regime (Corollary 8.3).

2. The phase field model and the macroscopic parameters

The difference between the previously studied phase field models [7 and the reference therein] and (1.8), (1.9) is the form of the entropy term arising from $-2u\gamma f(\varphi)$ in the free energy (1.6). The original equations in which $\gamma f(\varphi) = \varphi$ assume a linear approximation to the change in entropy density between phases. Although the linear approximation is convenient for many mathematical purposes, it is possible to consider nonlinear approximations within the transition region [12]. The physical accuracy can be expected to be of the same order as the linear approximation if the value of γ is adjusted to reflect the (macroscopic) difference in entropy density of the pure phases. (Other modified phase field models were recently reformulated in [12][19].)

Noting that the entropy difference incorporates temperature units, which in this case have been scaled by l_m/c_{spm}, one has the thermodynamic identity for the entropy difference per unit volume,

$$[s]_E = -\frac{\partial F}{\partial u}\bigg|_{\varphi=1} + \frac{\partial F}{\partial u}\bigg|_{\varphi=-1} = 2\gamma f(1) - 2\gamma f(-1) = 2\gamma \int_{-1}^{1} f'(\varphi)d\varphi. \tag{2.1}$$

Hence, the relation (2.1) *defines* γ in terms of a macroscopic observable.

Two scales emerge naturally from the phase equation (1.8): a length scale, ε, and a surface tension scale (Energy/Length^{N-1}), $\overline{\sigma}$, given by

$$\varepsilon := \xi a^{\frac{1}{2}} \qquad \text{and} \qquad \overline{\sigma} := \xi a^{-\frac{1}{2}}. \tag{2.2}$$

Noting that ε^2 is the coefficient of the Laplacian in (1.8), we define a coordinate system in which r is signed distance (positive in liquid) from the interface, $\Gamma(t)$, and a stretched or "inner" coordinate

$$\rho := r/\varepsilon. \tag{2.3}$$

If we define Φ as the solution of the ordinary differential equation

$$\Phi'' + g(\Phi) = 0, \qquad \Phi(\pm\infty) = \pm 1, \qquad \Phi(0) = 0, \tag{2.4}$$

then $\Phi(\rho)$ is the $O(1)$ inner expansion for φ (see [7] for more details).

We now focus on the surface tension, σ, in terms of its relation with ξ and a. Within a simply thermodynamic setting, the surface tension is obtained from a suitable local interpretation of the difference between the free energy of the system minus the average of the free energies in the pure phases (normalized with respect to surface area Λ); i.e.,

$$\sigma = \lim_{\text{measure of } \Lambda \to 0} \frac{\mathcal{F}_\Lambda(\Phi) - \frac{1}{2}(\mathcal{F}_\Lambda(+1) + \mathcal{F}_\Lambda(-1))}{\text{measure of } \Lambda}. \tag{2.5}$$

where F_Λ is defined by (1.6) evaluated in the domain Λ. A calculation [8; p239] shows that, to leading order in ε, the surface tension is given by

$$\sigma = \|\Phi'\|^2_{L^2(\mathbf{R})} \xi a^{-\frac{1}{2}} = m\overline{\sigma}, \qquad m := \|\Phi'\|^2_{L^2(\mathbf{R})}. \tag{2.6}$$

Noting that $a = \varepsilon/\overline{\sigma} = \varepsilon m/\sigma$, one can write (1.8) as

$$\alpha\varepsilon^2\varphi_t = \varepsilon^2\Delta\varphi + g(\varphi) + 2\frac{\gamma m\varepsilon}{\sigma}uf'(\varphi). \tag{2.7}$$

Using (2.1) for γ and defining

$$n := \frac{\|\Phi'\|^2_{L^2(\mathbf{R})}}{\int_{-1}^1 f'(\varphi)d\varphi} = \frac{m}{f(1)-f(-1)}, \tag{2.8}$$

we obtain finally the system

$$\alpha\varepsilon^2\varphi_t = \varepsilon^2\Delta\varphi + g(\varphi) + \frac{n}{d_0}\varepsilon uf'(\varphi), \tag{2.9}$$

$$u_t + \frac{1}{2}\varphi_t = D\Delta u \tag{2.10}$$

where $d_0 = \sigma/[s]_E$ since the additional factor has been absorbed into the dimensionless temperature.

For the prototype double-well potential

$$G(\varphi) = \frac{1}{8}(\varphi^2-1)^2, \qquad g(\varphi) = G'(\varphi) = \frac{1}{2}(\varphi - \varphi^3), \tag{2.11}$$

one has $m = 2/3$, while the choice of

$$f'(\varphi) = (1-\varphi^2)^2 \tag{2.12}$$

implies $n = 5/4$. In this case the solution Φ to (2.4) is given by

$$\Phi(\rho) = \tanh\frac{\rho}{2}. \tag{2.13}$$

3. A formal asymptotic analysis

We perform a preliminary asymptotic analysis for the equations (2.9), (2.10) which will establish the heuristic basis for the convergence of the phase field equations to the sharp interface model (1.1′)–(1.3′). Sections 4–8 will then provide a rigorous proof of the convergence to this limit.

The basic strategy in attaining this limit is similar to that of Section IV of [6]. Using the scaling defined by (2.2), one considers the distinguished limits as $\varepsilon \to 0$ but $\overline{\sigma}$ is held fixed. (This corresponds to fixed d_0 in (2.9).)

Suppose that in (2.9), (2.10), φ varies much more rapidly across the interface (from -1 in solid to $+1$ in liquid) than does u, and that φ can be approximated by a function of the form $\phi((r-vt)/\varepsilon)$; that is, the independent derivative with respect to time is of high order. Then (2.9) can be written as

$$-\alpha\varepsilon v\phi_\rho = \phi_{\rho\rho} + \varepsilon\kappa\phi_\rho + \cdots + g(\phi) + \varepsilon\frac{n}{d_0}uf'(\phi) \tag{3.1}$$

where "$\cdots$" are terms of order ε^2.

We assume an expansion of the form $\phi = \phi^0 + \varepsilon\phi^1 + \cdots$. Then equating the $O(1)$ terms in (3.1) gives

$$\phi^0_{\rho\rho} + g(\phi^0) = 0. \tag{3.2}$$

For the prototype potentials given by (2.11), the solution is given by (2.13). The equation for the $O(\varepsilon)$ order terms in (3.1) is

$$\phi^1_{\rho\rho} + g'(\phi^0)\phi^1 = H := -\alpha v\phi^0_\rho - \kappa\phi^0_\rho - \frac{n}{d_0}uf'(\phi^0). \tag{3.3}$$

Noting that the derivative of the $O(1)$ solution, ϕ^0_ρ, satisfies the homogeneous equation corresponding to (3.3), one obtains the solvability condition

$$0 = (\phi^0_\rho, H) = \int_{-\infty}^{\infty} \phi^0_\rho\big[-\alpha v\phi^0_\rho - \kappa\phi^0_\rho - \frac{n}{d_0}uf'(\phi^0)\big]\,d\rho. \tag{3.4}$$

Under the assumption that u varies slowly near the interface, i.e., when $f'(\phi^0)$ is of significant order, one has, upon using (2.8), the simplification (to leading order)

$$\int_{-\infty}^{\infty} \frac{n}{d_0}uf'(\phi^0)\phi^0_\rho d\rho = \frac{n}{d_0}u\int_{-1}^{1} f'(\phi^0)d\phi^0 = \frac{m}{d_0}u. \tag{3.5}$$

Finally, using the definition of m in (2.6), one obtains the interfacial relation (to leading order)

$$u = -d_0(\alpha v + \kappa) \qquad \text{on } \Gamma. \tag{3.6}$$

Note that ϕ^0 has a transition layer behavior at the interface and attains constant values outside of a region of width ε form the interface (similar to the function $\tanh r/2\varepsilon$ in the original coordinates), so that φ_t vanishes. Then an asymptotic solution (u, φ) to (2.9), (2.10) must be governed, to leading order, by a solution to the heat equation (1.1′) on $\Omega \setminus \Gamma$.

The latent heat condition (1.2′) is obtained by integrating (2.10) across the interface,

$$\int_{-\delta}^{\delta}\Big(u_t + \frac{1}{2}\varphi_t\Big)dr = \int_{-\delta}^{\delta}\Big(Du_{rr} + O(\varepsilon)\Big)dr, \tag{3.7}$$

so that the boundedness of u_t and the approximation $\varphi_t = -v\varphi_r$ implies the relation

$$v = D\big[u_r\big]^-_+ \tag{3.8}$$

which is equivalent to (1.2′).

Hence, a solution to the generalized phase field equation (2.9), (2.10) is expected to have formal asymptotics which are governed, to leading order, by the sharp interface problem (1.1′)–(1.3′).

4. Statement of the rigorous result

In this and subsequent sections we present the rigorous proof of the assertions made as a result of the formal asymptotics in Section 3. It is convenient to replace, without loss of generality, the coefficients α, n/d_0, and D, in (2.9), (2.10) by unity since the constants do not influence the proof in a significant way. Also, we use the prototype $g(\varphi) = 2(\varphi - \varphi^3)$ since the general case is very similar. We then consider the system of equations

$$\varphi_t - \Delta\varphi = \frac{1}{\varepsilon^2}(1-\varphi^2)\{2\varphi + \varepsilon u(1-\varphi^2)\}, \tag{4.1}$$

$$u_t - \Delta u = -\frac{1}{2}\varphi_t \tag{4.2}$$

in $Q_T \equiv \Omega \times (0,T)$, $\Omega \equiv \{x \in \mathbb{R}^N : r_1 < |x| < r_2\}$, subject to the initial–boundary conditions

$$u(x,t) = u_0(x,t), \qquad (x,t) \in \partial_T Q := \Omega \times \{0\} \cup \partial\Omega \times [0,T], \tag{4.3}$$

$$\varphi(x,0) = \Psi\left(\frac{|x| - r_0}{\varepsilon}\right), \quad x \in \Omega, \tag{4.4}$$

$$\varphi_r = \frac{1}{\varepsilon}(1-\varphi^2), \qquad (x,t) \in \partial\Omega \times [0,T] \tag{4.5}$$

where r_1, r_0 and r_2 are three given constants satisfying $r_1 < r_0 < r_2$, and

$$\Psi(\rho) = \tanh\rho, \quad -\infty < \rho < \infty. \tag{4.6}$$

Although more general boundary conditions for u can be used without significant changes in the proofs, we shall use (4.3) for definiteness.

The initial condition for φ ensures that the initial shape is compatible with the basic length scale in the problem [7, 8]. The boundary condition for φ is compatible with the initial condition and with vanishing φ_t and φ_x near the external boundary which are necessary in order to attain the heat equation in the limit. While other conditions may also be used, this particular condition is technically convenient.

For any $1 \le p \le \infty$, $\alpha \in (0,1)$, and $Q_t := \Omega \times (0,t)$, we introduce the norms:

$$\|f\|_{W_p^{2,1}(Q_t)} := \sum_{m+2k\le 2} \|\partial_x^m \partial_t^k f\|_{L^p(Q_t)}, \tag{4.7}$$

$$\|f\|_{C^\alpha(Q_t)} := \|f\|_{C^0(Q_t)} + \sup_{\substack{(x_1,t_1),\,(x_2,t_2)\in Q_t \\ (x_1,t_1)\neq(x_2,t_2)}} \frac{|f(x_1,t_1) - f(x_2,t_2)|}{|x_1 - x_2|^\alpha + |t_1 - t_2|^\alpha}, \tag{4.8}$$

$$\|f\|_{C^{\alpha,\alpha/2}(Q_t)} := \|f\|_{C^0(Q_t)} + \sup_{\substack{(x_1,t_1),\,(x_2,t_2)\in Q_t \\ (x_1,t_1)\neq(x_2,t_2)}} \frac{|f(x_1,t_1) - f(x_2,t_2)|}{|x_1 - x_2|^\alpha + |t_1 - t_2|^{\alpha/2}}, \tag{4.9}$$

$$\|f\|_{C^{1+\alpha,(1+\alpha)/2}(Q_t)} := \|f\|_{C^{(1+\alpha)/2}(Q_t)} + \|\partial_x f\|_{C^{\alpha,\alpha/2}(Q_t)}. \tag{4.10}$$

We state a standard result concerning the system (4.1)–(4.5).

LEMMA 4.1. *Assume that* $u_0 \in C^{2,1}(Q_T)$. *Then for every* $\varepsilon > 0$, *the system (4.1)–(4.5) has a unique classical solution* $(u^\varepsilon, \varphi^\varepsilon)$. *Moreover,*

$$|\varphi^\varepsilon(x,t)| < 1 \qquad \forall (x,t) \in \overline{Q_T} \equiv \overline{\Omega} \times [0,T]. \tag{4.11}$$

The proof of the existence of a unique solution is similar to that in [8], whereas (4.11) follows by applying the maximum principle to the parabolic equation (4.1) (treating u^ε as a known coefficient).

From now on we shall prove the following theorem:

THEOREM 4.1. *Assume that* $u_0 \in C^{2,1}(Q_T)$ *and is radially symmetric, and let* $(u^\varepsilon, \varphi^\varepsilon)$ *be the solution of (4.1)–(4.5). Then there exist functions* $u(x,t) \in C^{\alpha,\alpha/2}(Q_{T^*})$ *and* $S(t) \in C^{1+\alpha/2}([0,T^*))$, *such that*

$$\lim_{\varepsilon \to 0+} u^\varepsilon(x,t) = u(x,t), \qquad \forall (x,t) \in \overline{\Omega} \times [0,T^*), \tag{4.12}$$

$$\lim_{\varepsilon \to 0+} \varphi^\varepsilon(x,t) = \begin{cases} 1 & \text{if } S(t) < |x| \le r_2,\ 0 \le t < T^*, \\ -1 & \text{if } r_1 \le |x| < S(t),\ 0 \le t < T^* \end{cases} \tag{4.13}$$

where $T^* > 0$ *is the first time such that one of the following happens:*

$$T^* = T; \qquad S(T^*) = r_2; \qquad S(T^*) = r_1. \tag{4.14}$$

Moreover, if we denote by Γ *the set* $\{(x,t) \in Q_{T^*} : |x| = S(t)\}$, *then* (u, Γ) *is a solution to* (1.1′)–(1.3′); *that is,*

$$u_t = \Delta u \qquad \text{in } Q_{T^*} \setminus \Gamma, \tag{4.15}$$

$$\dot{S}(t) = [u_r(x,t)]_+^- \qquad \text{on } \Gamma, \tag{4.16}$$

$$\dot{S}(t) = -\frac{N-1}{S(t)} - \beta u(S(t),t), \qquad \text{on } [0,T^*) \tag{4.17}$$

where β *is a positive constant defined in Section 8 below.*

Recall that the sum of principle curvatures of a ball of radius r is $\frac{N-1}{r}$, so that equation (4.17) is equivalent to (1.3′).

To explain the idea of the proof of the theorem, we introduce a function $Z^\varepsilon(x,t) : Q_T \to \mathbf{R}^1$ defined by

$$Z^\varepsilon(x,t) = \varepsilon \Psi^{-1}\big(\varphi^\varepsilon(x,t)\big). \tag{4.18}$$

Since (4.11) implies that the value of φ^ε is in the range of Ψ, the function Z^ε is well–defined. Clearly the definition of Z^ε implies that

$$\varphi^\varepsilon(x,t) = \Psi\Big(\frac{Z^\varepsilon(x,t)}{\varepsilon}\Big), \qquad (x,t) \in Q_T. \tag{4.19}$$

The overall strategy for the proof of the theorem is to show that Z^ε is approximately equal to $|x| - S(t)$ for some function $S(t) \in C^\alpha$, $\alpha \in (1/2, 1)$. Once we prove this, we can substitute (4.19) into (4.2) to obtain Hölder estimates, for u^ε, independent of ε, by applying a potential analysis to the Green's representation for the solution u^ε of the heat equation (4.2). Having known the Hölder continuity of the function u^ε, we then go back to the equation (4.1), which has been well-studied in the case u^ε being a constant [2, 5, 13, 23, 24, 25], a known function [18], or an unknown function satisfying a parabolic equation coupled with φ^ε [14, 16, 26]. The conclusion is expressed by equations (4.13) and (4.17). Finally, by using (4.13) and the distribution sense of equation (4.2), one obtains (4.15) and (4.16).

To prove that Z^ε is approximately equal to $|x| - S(t)$, we need, however, some regularity on u^ε. For this reason, we introduce the functions:

$$M_\varepsilon^0(t) := \|u^\varepsilon\|_{C^0(Q_t)}, \tag{4.20}$$

$$M_\varepsilon^1(t) := \|u_r^\varepsilon\|_{C^0(Q_t)}, \tag{4.21}$$

$$M_\varepsilon(t) := M_\varepsilon^0(t) + \varepsilon M_\varepsilon^1(t), \tag{4.22}$$

$$T^\varepsilon := \sup\{t \in [0, T] : M_\varepsilon(\tau) \le \frac{1}{\sqrt{\varepsilon}} \text{ for all } \tau \in [0, t]\}. \tag{4.23}$$

By restricting oneself to the interval $[0, T^\varepsilon]$, one can carry out all the steps described in the proceeding paragraph. Therefore, to complete the proof, one need only show that the a priori estimate thus obtained in the interval $[0, T^\varepsilon]$ is independent of ε since this means that $T^\varepsilon = T$.

In the following we shall denote by C the various kinds of constants which are independent of ε. Also, we shall identify functions of variable (x, t) with functions of variable (r, t) with $r = |x|$ since all function in the sequel are radially symmetric. Finally, we shall assume, without loss of generality, that

$$M_\varepsilon(t) \ge 1, \qquad t \in [0, T]. \tag{4.24}$$

5. $C^{(1+\alpha)/2}$ estimate for the interface

In this section, we shall show that the interface which coincides with the zero level set of Z^ε is $C^{(1+\alpha)/2}$ for any $\alpha \in (0, 1)$.

LEMMA 5.1. *Let Z^ε and T^ε be defined as in (4.18) and (4.23). Then there exist positive constants ε_0 and $C > 0$ such that for all $\varepsilon \in (0, \varepsilon_0)$ and $(x, t) \in \overline{Q_{T^\varepsilon}}$, one has*

$$|Z^\varepsilon(x, t)| \le 2r_2 + \frac{N}{r_1} T^\varepsilon, \tag{5.1}$$

$$\frac{1}{2} \le Z_r^\varepsilon(x, t) \le 2, \tag{5.2}$$

$$-C\varepsilon M_\varepsilon(t) \le Z_r^\varepsilon(x, t) - 1 \le C\varepsilon M_\varepsilon(t). \tag{5.3}$$

Note that (5.3) is stronger than (5.2).

Proof. Substituting (4.19) into (4.1) yields the equation

$$(5.4)\qquad \Big(Z^\varepsilon_t - \Delta Z^\varepsilon\Big)\Psi' - \frac{|\nabla Z^\varepsilon|^2}{\varepsilon}\Psi'' - \frac{1}{\varepsilon}(1-\Psi^2)\Big(2\Psi + \varepsilon u^\varepsilon(1-\Psi^2)\Big) = 0.$$

Using the radial coordinates and the identities

$$\Psi' = 1 - \Psi^2, \qquad \Psi'' = -2\Psi\Psi',$$

one can write (5.4) as

$$(5.5)\qquad Z^\varepsilon{}_t - Z^\varepsilon{}_{rr} - \frac{N-1}{r}Z^\varepsilon{}_r + \frac{2}{\varepsilon}(Z^{\varepsilon 2}_r - 1)\Psi\Big(\frac{Z^\varepsilon}{\varepsilon}\Big) - u^\varepsilon\Psi'\Big(\frac{Z^\varepsilon}{\varepsilon}\Big) = 0,$$

and write the initial–boundary conditions (4.4), (4.5) as

$$(5.6)\qquad Z^\varepsilon(r,0) = r - r_0, \qquad r \in (r_1, r_2),$$

$$(5.7)\qquad Z^\varepsilon{}_r(r_1,t) = Z^\varepsilon{}_r(r_2,t) = 1, \qquad t \in [0,T]$$

where we have identified the function $Z^\varepsilon(x,t)$ with the function $Z^\varepsilon(r,t)$ with $r = |x|$.

Noting that

$$\Psi'\Big(\frac{Z^\varepsilon}{\varepsilon}\Big) = \cosh^{-2}\Big(\frac{Z^\varepsilon}{\varepsilon}\Big) \le 4e^{-\frac{|Z^\varepsilon|}{\varepsilon}} \le 4\varepsilon^k \quad \text{if } |Z^\varepsilon| \ge k\varepsilon|\ln\varepsilon|,$$

one can directly verify that for ε sufficient small, the functions

$$Z^+ := r + \frac{N}{r_1}t$$

and

$$Z^- := r - 2r_2 - \varepsilon t$$

are, respectively, a supersolution and a subsolution to (5.5)–(5.7), so that

$$Z^-(r,t) \le Z^\varepsilon(r,t) \le Z^+(r,t), \qquad r_1 \le r \le r_2,\ 0 \le t \le T^\varepsilon,$$

and therefore the asssertion (5.1) follows.

To prove (5.2), we differentiate (5.5) with respect to r and set $Z^\varepsilon{}_r = w$, obtaining a nonlinear parabolic equation, for w,

$$\begin{aligned}\mathcal{N}w := w_t - w_{rr} - \frac{N-1}{r}w_r + \frac{N-1}{r^2}w + \frac{4}{\varepsilon}\Psi w w_r \\ (5.8)\qquad + \frac{2}{\varepsilon^2}\Psi'(w^2-1)w - u^\varepsilon_r\Psi' + \frac{2}{\varepsilon}u^\varepsilon\Psi\Psi' w = 0, \qquad (x,t) \in Q_T,\end{aligned}$$

$$(5.9)\qquad w = 1, \qquad (x,t) \in \partial_T Q.$$

Set

$$w^+(r,t) := 1 + 2\varepsilon M_\varepsilon(\bar t), \qquad r \in [r_1, r_2],\ 0 \le t \le \bar t \le T^\varepsilon$$

and

$$w^-(r,t) := 1 - 2\varepsilon M_\varepsilon(\bar{t}) - \frac{N-1}{r_1^2}t, \qquad r \in [r_1, r_2],\ 0 \le t \le \bar{t} \le T_1^\varepsilon$$

where $T_1^\varepsilon \in (0, T^\varepsilon]$ is any constant which can ensure $w^- \ge \frac{1}{2}$ for all $t \in [0, T_1^\varepsilon]$. One can verify, by using the definition of M_ε and the fact that $\frac{1}{2} < w^- < w^+ < 2$, that w^+ and w^- are, respectively, a supersolution and a subsolution of (5.8), (5.9). Therefore, a comparison principle for parabolic equation implies that $w^- \le {Z^\varepsilon}_r \le w^+$, which, in turn, implies that (5.2) is valid in $[0, T_1^\varepsilon]$.

We shall now use (5.2) to prove (5.3). Clearly, we need only prove the first inequality in (5.3) since the inequality $w \le w^+$ implies the second inequality in (5.3).

We claim that, for a suitable pair of constants k_1 and k_2 which are independent of ε, the function

$$\underline{w} := 1 - k_1\varepsilon\sqrt{{Z^\varepsilon}^2 + k_1^2\varepsilon^2} - k_2\varepsilon M_\varepsilon \tag{5.10}$$

is a subsolution to (5.8), (5.9).

One can compute

$$\begin{aligned}
I :=& \underline{w}_t - \underline{w}_{rr} - \frac{N-1}{r}\underline{w}_r + \frac{N-1}{r^2}\underline{w} \\
=& -\frac{k_1\varepsilon Z^\varepsilon}{({Z^\varepsilon}^2 + k_1^2\varepsilon^2)^{1/2}}\Big({Z^\varepsilon}_t - {Z^\varepsilon}_{rr} - \frac{N-1}{r}{Z^\varepsilon}_r\Big) + \frac{k_1^3\varepsilon^3}{({Z^\varepsilon}^2 + k_1^2\varepsilon^2)^{3/2}}{Z^\varepsilon}_r^2 + \frac{N-1}{r^2}\underline{w} \\
=& -\frac{k_1\varepsilon Z^\varepsilon}{({Z^\varepsilon}^2 + k_1^2\varepsilon^2)^{1/2}}\Big(-\frac{2}{\varepsilon}({Z^\varepsilon}_r^2 - 1)\Psi + u^\varepsilon\Psi'\Big) + \frac{k_1^3\varepsilon^3}{({Z^\varepsilon}^2 + k_1^2\varepsilon^2)^{3/2}}{Z^\varepsilon}_r^2 + \frac{N-1}{r^2}\underline{w} \\
\le& 2k_1 \max\{{Z^\varepsilon}_r^2 - 1, 0\}|\Psi| + k_1\varepsilon|u^\varepsilon|\Psi' + 4 + \frac{N-1}{r^2} \\
\le& k_1\varepsilon M_\varepsilon(8|\Psi| + \Psi') + 4 + \frac{N-1}{r^2},
\end{aligned}$$

where in the third equation, we have used equation (5.5), in the first inequality, we have used the fact that $Z^\varepsilon\Psi \ge 0$, $\frac{Z^\varepsilon}{({Z^\varepsilon}^2 + k_1^2\varepsilon^2)^{\frac{1}{2}}} \le 1$, and ${Z^\varepsilon}_r^2 \le 4$, and in the last inequality, we have used the fact that ${Z^\varepsilon}_r^2 - 1 \le {w^+}^2 - 1 \le 4\varepsilon M_\varepsilon$ and $|u^\varepsilon| \le M_\varepsilon$. One can also compute

$$\begin{aligned}
II :=& \frac{4}{\varepsilon}\Psi\underline{w}\underline{w}_r + \frac{2}{\varepsilon^2}\Psi'(\underline{w}^2 - 1)\underline{w} \\
=& -4k_1\Psi\frac{\underline{w}Z^\varepsilon {Z^\varepsilon}_r}{({Z^\varepsilon}^2 + k_1^2\varepsilon^2)^{1/2}} + \frac{2}{\varepsilon^2}\Psi'(\underline{w}+1)(\underline{w}-1)\underline{w} \\
\le& -k_1|\Psi|\chi_{\{|Z^\varepsilon| \ge k_1\varepsilon\}} - \frac{2k_2M_\varepsilon}{\varepsilon}\Psi'
\end{aligned}$$

since ${Z^\varepsilon}_r > 1/2$. Finally, one has

$$III := -u_r^\varepsilon\Psi' + \frac{2}{\varepsilon}u^\varepsilon\Psi\Psi'\underline{w} \le 3\frac{M_\varepsilon}{\varepsilon}\Psi'.$$

Therefore, the sum satisfies the inequality

$$\begin{aligned}\mathcal{N}\underline{w} &= I + II + III \\ &\leq k_1 \varepsilon M_\varepsilon (8|\Psi| + \Psi') + 4 + \frac{N-1}{r^2} - k_1|\Psi|\chi_{\{|Z^\varepsilon| \geq k_1\varepsilon\}} - \frac{2k_2 M_\varepsilon}{\varepsilon}\Psi' + 3\frac{M_\varepsilon}{\varepsilon}\Psi'. \\ &\leq C k_1 \sqrt{\varepsilon} + C - k_1 \min\left\{\Psi(k_1), \frac{k_2 M_\varepsilon}{k_1 \varepsilon}\Psi'(k_1)\right\} \leq 0\end{aligned}$$

if we first take k_1 large enough, and then k_2 large enough and ε small enough. Therefore by comparison, $\underline{w} \leq {Z^\varepsilon}_r$, which implies that the first inequality in (5.3) holds for $t \in [0, T_1^\varepsilon]$.

Repeating the above proof in the interval $[T_1^\varepsilon, T^\varepsilon]$, one can easily extend, step by step, the valid interval for (5.2), (5.3) up to $[0, T^\varepsilon]$, and therefore complete the proof of the lemma. □

We can now obtain L^p and Hölder estimates for Z^ε (based on M_ε).

LEMMA 5.2. *For all $p \in (1, \infty)$ and $\alpha \in (0,1)$, there exist constants C_p and C_α, which are independent of ε, such that for all $t \in [0, T^\varepsilon]$ one has the bounds*

$$\|Z^\varepsilon\|_{W_p^{2,1}(Q_t)} \leq C_p M_\varepsilon(t), \tag{5.11}$$

$$\|Z^\varepsilon\|_{C^{1+\alpha,(1+\alpha)/2}(Q_t)} \leq C_\alpha M_\varepsilon(t). \tag{5.12}$$

Proof. The equation (5.5), along with the inequality (5.3), implies the inequalities

$$|{Z^\varepsilon}_t - {Z^\varepsilon}_{rr}| \leq \frac{N-1}{r}|{Z^\varepsilon}_r| + 2|{Z^\varepsilon}_r + 1|\frac{|{Z^\varepsilon}_r - 1|}{\varepsilon}|\Psi| + |u^\varepsilon|\Psi' \leq C M_\varepsilon(t).$$

Then (5.11) follows from the classical L^p estimates while (5.12) follows from Sobolev embedding Theorem. □

Since Z^ε is strictly increasing (by (5.2)), one can define the inverse, $r = \widetilde{R^\varepsilon}(z,t)$, of the function $z = Z^\varepsilon(r,t)$. It is convenient to extend $\widetilde{R^\varepsilon}$ to R^ε on $\mathbb{R}^+ \times [0, T^\epsilon]$ defined by

$$R^\varepsilon(z,t) := \begin{cases} \widetilde{R^\varepsilon}(z,t) & \text{if } z \in [Z^\varepsilon(r_1,t), Z^\varepsilon(r_2,t)], \\ r_1 & \text{if } z < Z^\varepsilon(r_1,t), \\ r_2 & \text{if } z > Z^\varepsilon(r_2,t). \end{cases} \tag{5.13}$$

The estimate (5.3) then implies that

$$Z^\varepsilon(r,t) = \left[r - R^\varepsilon(0,t)\right]\left[1 + O(\varepsilon M_\varepsilon)\right]. \tag{5.14}$$

LEMMA 5.3. *For all $\alpha \in (0,1)$ there exists a constant $C_\alpha > 0$ such that*

$$\|R^\varepsilon\|_{C^\alpha(Q_t)} \leq C_\alpha M_\varepsilon(t) \qquad \forall t \in (0, T^\varepsilon]. \tag{5.15}$$

This lemma follows from Lemma 5.2 and the estimate (5.2).

6. A L^∞ bound on u^ε using Green's Function

We shall now use the heat equation (4.2) to estimate the L^∞ bound for u^ε.

Let $G(x,t;\xi,\tau)$ be the Green's function corresponding to the boundary conditions imposed on u; that is, G satisfies

$$G_\tau + \Delta_\xi G = 0, \qquad (x,\xi)\in\Omega\times\Omega,\ 0\le\tau<t, \tag{6.1}$$

$$G(x,t;\xi,\tau) = 0, \qquad (x,t)\in Q_T,\ (\xi,\tau)\in\partial\Omega\times[0,T), \tag{6.2}$$

$$\lim_{\tau\to t^-} G(x,t;\xi,\tau) = \delta(x-\xi), \qquad (x,\xi,t)\in\Omega\times\Omega\times(0,T] \tag{6.3}$$

where the δ function in the last equation has the standard interpretation in the sense of distributions. By Green's formula, a solution u^ε to (4.2), (4.3) is given by

$$\begin{aligned} u^\varepsilon(x,t) &= \int_\Omega G(x,t;\xi,0)u_0(x,0)d\xi - \int_0^t\int_{\partial\Omega}\frac{\partial G}{\partial_\xi n}(x,t;\xi,\tau)u_0(\xi,\tau)dS_\xi d\tau \\ &\quad -\frac{1}{2}\int_0^t\int_\Omega G(x,t,\xi,\tau)\varphi^\varepsilon_\tau(\xi,\tau)d\xi d\tau \\ &\equiv u_1(x,t)+u_2(x,t)-\frac{1}{2}u_3(x,t) \end{aligned} \tag{6.4}$$

where n is the normal to the surface element dS of $\partial\Omega$.

Note that u_1+u_2 is a solution to

$$v_t - \Delta v = 0 \quad \text{in } Q_T, \tag{6.5}$$

$$v = u_0 \quad \text{on } \partial_T Q, \tag{6.6}$$

so that one has, for a constant M_0 (depending only on g, Ω, and T), the bound

$$\|u_1+u_2\|_{C^{1,1/2}(Q_T)} \le M_0. \tag{6.7}$$

Hence, we need only analyze the regularity of u_3 in order to obtain the L^∞ or Hölder estimate for u^ε.

The following lemma establishes a recursive relation for $M^1_\varepsilon(t)$.

LEMMA 6.1. *For any $\alpha\in(0,1)$ and $q\in(1,\frac{3}{2+\alpha})$, there exists a constant $C_{\alpha,q}$, which is independent of ε, such that for all $t\in[0,T^\varepsilon]$,*

$$M^1_\varepsilon := \left\|\frac{\partial u^\varepsilon}{\partial r}\right\|_{C^0(Q_t)} \le C_{\alpha,q}\varepsilon^{\alpha-1}M_\varepsilon(t)t^{\frac{3-(\alpha+2)q}{2q}} + M_0. \tag{6.8}$$

Proof. We need only estimate $|u_{3,r}|$. Denoting

$$\widetilde{G}(r,t;r',\tau) := \int_{\{|\xi|=r'\}} G(x,t;\xi,\tau)\Big|_{|x|=r} dS_\xi d\tau, \tag{6.9}$$

one gets

$$u_3(r,t) = \int_0^t \int_{r_1}^{r_2} \widetilde{G}(r,t,r',\tau)\varphi^\varepsilon_\tau(r',\tau)dr'd\tau = \frac{1}{\varepsilon}\int_o^t \int_{r_1}^{r_2} \widetilde{G}\Psi'\Big(\frac{Z^\varepsilon}{\varepsilon}\Big)Z^\varepsilon{}_\tau dr'd\tau.$$

Differentiating both sides with respect to r yields

$$\Big|\frac{\partial u_3}{\partial r}\Big| \leq \frac{1}{\varepsilon}\int_0^t \int_{r_1}^{r_2} \Big|\frac{\partial \widetilde{G}}{\partial r}\Big|\Psi'\Big(\frac{Z^\varepsilon}{\varepsilon}\Big)|Z^\varepsilon{}_\tau|dr'd\tau$$

$$\leq \frac{1}{\varepsilon}\|Z^\varepsilon{}_\tau\|_{L^{q'}}\Big\{\int_0^t \int_{r_1}^{r_2} |\widetilde{G}_r|^q|\Psi'|^q dr'd\tau\Big\}^{1/q} \tag{6.10}$$

$$\leq \frac{1}{\varepsilon}\|Z^\varepsilon{}_\tau\|_{L^{q'}}\Big\{\Big(\sup_{\tau\in[0,t]}\int_{r_1}^{r_2} |\Psi'|^{pq}\Big)^{1/p}\int_0^t \Big(\int_{r_1}^{r_2} \widetilde{G}_r|^{qp'}\Big)^{1/p'}\Big\}^{1/q}$$

where $1/q + 1/q' = 1$ ($q > 1$) and $1/p + 1/p' = 1$ ($p > 1$).

One may write a basic inequality [20, Chapter 1] involving the Green's function as

$$\int_0^t \Big(\int_{r_1}^{r_2} |\widetilde{G}_r|^{qp'}dr'\Big)^{1/p'} d\tau \leq C\int_0^t \Big\{\int_{r_1}^{r_2} \Big(1+\frac{|r-r'|}{|t-\tau|^{3/2}}e^{-\frac{|r-r'|^2}{4|t-\tau|}}\Big)^{qp'} dr'\Big\}^{1/p'}$$

$$\leq Ct^{\frac{1}{2p}+1-q} \tag{6.11}$$

where C depends only on Ω. Since $0 < \Psi' < 1$ and $Z^\varepsilon{}_r > 1/2$, one has

$$\int_{r_1}^{r_2} |\Psi'|^{pq} \leq \int_{r_1}^{r_2} \Psi' = \varepsilon\int_{Z^\varepsilon(r_1,\tau)}^{Z^\varepsilon(r_2,\tau)} \Psi'\Big(\frac{Z^\varepsilon}{\varepsilon}\Big)\frac{1}{Z^\varepsilon{}_r}d\Big(\frac{Z^\varepsilon}{\varepsilon}\Big) \leq 4\varepsilon. \tag{6.12}$$

Setting $p = \frac{1}{\alpha q}$, substituting (6.11), (6.12) into (6.10), and using the L^p estimate for Z^ε [(5.11)], one obtains the bound

$$\Big|\frac{\partial u_3}{\partial r}\Big| \leq C\frac{1}{\varepsilon}M_\varepsilon(t)t^{[\frac{1}{2p'}+1-q]/q}\varepsilon^\alpha = C\varepsilon^{\alpha-1}M_\varepsilon(t)t^{\frac{3-(\alpha+2)q}{2q}},$$

and therefore the lemma follows. □

To get the L^∞ bound for u_3, we utilize the identities

$$u_3(x,t) = \int_0^t \int_\Omega G(x,t;\xi,\tau)\frac{\partial}{\partial\tau}\Big[\varphi^\varepsilon(\xi,\tau)-\varphi^\varepsilon(\xi,t)\Big]d\xi d\tau$$

$$= \int_\Omega G\Big[\varphi^\varepsilon(\xi,\tau)-\varphi^\varepsilon(\xi,t)\Big]\Big|_{\tau=0}^{\tau=t} d\xi - \int_0^t \int_\Omega G_\tau\Big[\varphi^\varepsilon(\xi,\tau)-\varphi^\varepsilon(\xi,t)\Big]d\xi d\tau$$

$$= \int_\Omega G(x,t;\xi,0)\Big[\varphi^\varepsilon(\xi,t)-\varphi^\varepsilon(\xi,0)\Big]d\xi + \int_0^t \int_\Omega \Delta_\xi G\Big[\varphi^\varepsilon(\xi,\tau)-\varphi^\varepsilon(\xi,t)\Big]d\xi d\tau$$

$$= \int_\Omega G(x,t;\xi,0)\Big[\varphi^\varepsilon(\xi,t)-\varphi^\varepsilon(\xi,0)\Big]d\xi$$

$$+ \int_0^t \int_{\partial\Omega} \frac{\partial G}{\partial r'}(x,t;\xi,\tau)\Big[\varphi^\varepsilon(\xi,\tau)-\varphi^\varepsilon(\xi,t)\Big]d\xi d\tau$$

$$- \int_0^t \int_\Omega \frac{\partial G}{\partial r'}(x,t;\xi,\tau)\Big[\frac{\partial\varphi^\varepsilon}{\partial r'}(\xi,\tau)-\frac{\partial\varphi^\varepsilon}{\partial r'}(\xi,t)\Big]d\xi d\tau$$

$$\equiv A(x,t)+B(x,t)+C(x,t) \tag{6.13}$$

where integration by parts in t, the heat equation for G, and Green's theorem have been used.

LEMMA 6.2. *For any $\alpha \in (1/2,1)$, there exists a positive constant C_α, such that for any positive constant δ and for all $(x,t) \in Q_{T^\varepsilon}$ one has the bounds*

$$|A(x,t)+B(x,t)| \leq 2, \tag{6.14}$$

$$|C(x,t)| \leq 8 + C_\alpha\Big(\delta^{-\frac{3}{2}} + M_\varepsilon(t)\delta^{\alpha-\frac{1}{2}}\Big), \tag{6.15}$$

$$M_\varepsilon^0 \leq C + C_\alpha\Big(\delta^{-\frac{3}{2}} + M_\varepsilon(t)\delta^{\alpha-\frac{1}{2}}\Big). \tag{6.16}$$

Proof. We need only prove (6.14) and (6.15) since (6.16) follows from (6.4), (6.7), (6.14) and (6.15).

Using the bound $|\varphi| < 1$ (Lemma 4.1), one obtains

$$|A+B| \leq 2\sup|\varphi|\Big\{\int_\Omega G(x,t;\xi,0)d\xi + \int_0^t\int_{\partial\Omega}\frac{\partial G}{\partial n}(x,t;\xi,\tau)dS_\xi d\tau\Big\} \leq 2.$$

Write $C(x,t)$ as

$$C(x,t) = \Big(\int_0^{\max\{0,t-\delta\}} + \int_{\max\{0,t-\delta\}}^t\Big)\int_\Omega \cdots d\xi d\tau = C^{(1)} + C^{(2)},$$

$$\tag{6.17} \begin{aligned} C^{(1)} &:= \int_0^{\max\{0,t-\delta\}}\int_{r_1}^{r_2} \widetilde{G}_{r'}(r,t;r',\tau)\Big[\varphi^\varepsilon_{r'}(r',t) - \varphi^\varepsilon_{r'}(r',\tau)\Big]dr'd\tau, \\ C^{(2)} &:= \int_{\max\{0,t-\delta\}}^t\int_{r_1}^{r_2} \widetilde{G}_{r'}(r,t;r',\tau)\Big[\varphi^\varepsilon_{r'}(r',t) - \varphi^\varepsilon_{r'}(r',\tau)\Big]dr'd\tau. \end{aligned}$$

Integrating by parts for the integral $C^{(1)}$ and using the bound

$$|\widetilde{G}_{r'r'}(r,t,r',\tau)| \leq C(t-\tau)^{-\frac{3}{2}},$$

one finds that $C^{(1)}$ is bounded by $C\delta^{-\frac{3}{2}}$.

To estimate $C^{(2)}$, we substitute φ^ε by $\Psi(Z^\varepsilon/\varepsilon)$ in (6.17) and use the change of variables $\eta = Z^\varepsilon/\varepsilon$, obtaining

$$\begin{aligned} C^{(2)}(x,t) = &\int_{\max\{0,t-\delta\}}^t \int_{Z^\varepsilon(r_1,t)/\varepsilon}^{Z^\varepsilon(r_2,t)/\varepsilon} \widetilde{G}_{r'}(r,t;R^\varepsilon(\varepsilon\eta,t),\tau)\Psi'(\eta)d\eta d\tau \\ &- \int_{\max\{0,t-\delta\}}^t \int_{Z^\varepsilon(r_1,\tau)/\varepsilon}^{Z^\varepsilon(r_2,\tau)/\varepsilon} \widetilde{G}_{r'}(r,t;R^\varepsilon(\varepsilon\eta,\tau),\tau)\Psi'(\eta)d\eta d\tau. \end{aligned}$$

By dividing the η integration of the first integral into the three parts:

$$\int_{Z^\varepsilon(r_1,t)/\varepsilon}^{Z^\varepsilon(r_1,\tau)/\varepsilon} + \int_{Z^\varepsilon(r_1,\tau)/\varepsilon}^{Z^\varepsilon(r_2,\tau)} + \int_{Z^\varepsilon(r_2,\tau)/\varepsilon}^{Z^\varepsilon(r_2,t)/\varepsilon},$$

one obtains the bound

$$
\begin{aligned}
&|C^{(2)}(x,t)| \\
\leq &\int_{\max\{0,t-\delta\}}^{t} d\tau \int_{\frac{Z^\varepsilon(r_1,\tau)}{\varepsilon}}^{\frac{Z^\varepsilon(r_2,\tau)}{\varepsilon}} \left|\widetilde{G}_{r'}(r,t;R^\varepsilon(\varepsilon\eta,t),\tau) - \widetilde{G}_{r'}(r,t,R^\varepsilon(\varepsilon\eta,\tau),\tau)\right| \Psi'(\eta)d\eta \\
&\quad + 2\int_{-\infty}^{\infty} d\eta \Psi'(\eta) \int_{\max\{0,t-\delta,0\}}^{t} |\widetilde{G}_{r'}(r,t,R^\varepsilon(\varepsilon\eta,t),\tau)|d\tau \\
\equiv &C^{(21)} + C^{(22)}.
\end{aligned}
$$

The integral $C^{(22)}$ is bounded by 8 since for all $r, r' \in (r_1, r_2)$, one has

$$\int_{\max\{0,t-\delta\}}^{t} |\widetilde{G}_{r'}(r,t,r',\tau)|d\tau \leq 2.$$

The integral $C^{(21)}$ can be estimated by

$$
\begin{aligned}
C^{(21)} \leq &\int_{\max\{t-\delta,0\}}^{t} d\tau \int_{Z^\varepsilon(r_1,t)/\varepsilon}^{Z^\varepsilon(r_2,t)/\varepsilon} d\eta \Psi'(\eta) |\widetilde{G}_{r'r'}(r,t;\overline{r'},\tau)| R^\varepsilon(\varepsilon\eta,\tau) - R^\varepsilon(\varepsilon\eta,t)| \\
\leq &\int_{\max\{t-\delta,0\}}^{t} d\tau \int_{-\infty}^{\infty} d\eta \Psi'(\eta) C|t-\tau|^{-3/2} C_\alpha M_\varepsilon(t)|t-\tau|^\alpha \\
\leq &\widetilde{C}_\alpha M_\varepsilon(t)\delta^{\alpha-1/2}
\end{aligned}
$$

where the mean value theorem and Lemma 5.3 have been used. Combining all the estimates, one obtains (6.15) and the lemma. □

THEOREM 6.1. *There exist positive constants ε_0 and C such that for for all $\varepsilon \in (0, \varepsilon_0)$, one has the bound*

$$\|u^\varepsilon\|_{L^\infty(Q_T)} \leq C. \tag{6.18}$$

Proof. Using (6.8) (with $q = \frac{1}{2}(1 + \frac{3}{2+\alpha})$), (6.16), and the definition of M_ε in (4.22), one has

$$M_\varepsilon(t) \leq C_{\alpha,T}\Big[1 + \delta^{-3/2} + M_\varepsilon(t)\delta^{\alpha-1/2} + \varepsilon^\alpha M_\varepsilon(t)\Big]. \tag{6.19}$$

Choosing ε_0 and δ satisfying

$$\varepsilon_0^\alpha C_{\alpha,T} \leq \frac{1}{4}, \qquad C_{\alpha,T}\delta^{\alpha-1/2} \leq \frac{1}{4},$$

one has, from (6.19), the bound

$$M_\varepsilon(t) \leq 2C_{\alpha,T}\Big[1 + \delta^{-\frac{3}{2}}\Big] \equiv \widetilde{M}_0, \qquad \text{for all } \; t \in [0, T^\varepsilon], \quad \varepsilon \in (0, \varepsilon_0]. \tag{6.20}$$

Further choosing ε_0 sufficient small such that

$$\widetilde{M}_0 \leq \frac{1}{2\sqrt{\varepsilon_0}},$$

one concludes, from the definition of T^ε in (4.23) and the estimate (6.20), that

$$T^\varepsilon = T$$

if $\varepsilon \leq \varepsilon_0$. This completes the proof of Theorem 6.1. □

With the estimate (6.20), Lemmas 5.2 and 5.3 can be strengthened as follows:

THEOREM 6.2. *There exists a constant $\varepsilon_0 > 0$ such that for all $\varepsilon \in (0, \varepsilon_0)$, $\alpha \in (0,1)$, one has*

$$\|R^\varepsilon\|_{C^\alpha(\mathbf{R}^1\times[0,T])} \le C_{\alpha,T}, \tag{6.21}$$

$$\|Z^\varepsilon\|_{C^{1+\alpha,(1+\alpha)/2}(Q_T)} \le C_{\alpha,T}. \tag{6.22}$$

7. Hölder estimates for u^ε

Theorem 6.2 implies that the interface (determined by $Z^\varepsilon = 0$ or $r = R^\varepsilon(0,t)$) does not intersect the external walls ($r = r_2$ or $r = r_2$) for a certain amount of time; that is, for any sufficient small positive constant a, the constant T_a defined by

$$T_a := \sup\{t \in (0,T] : Z^\varepsilon(r_1,\tau) \le -a,\ Z^\varepsilon(r_2,\tau) \ge a,\ \forall\varepsilon \in (0,a^2], \tau \in [0,t]\} \tag{7.1}$$

is positive.

THEOREM 7.1. *For any $\alpha \in (0,1)$ and $a > 0$ sufficiently small, there exists a positive constant $C_{\alpha,a}$ such that, for all $\varepsilon \in (0, a^2]$, one has*

$$\|u^\varepsilon\|_{C^{\alpha,\alpha/2}(Q_{T_a})} \le C_{\alpha,a} \tag{7.2}$$

where T_a is as in (7.1).

Proof. The definition of T_a implies that φ^ε is exponentially close to ± 1 at the external walls; i.e.,

$$\varphi^\varepsilon(x,t)^2 = \Psi^2\Big(\frac{Z^\varepsilon(x,t)}{\varepsilon}\Big) = 1 + O(e^{-a/\varepsilon}), \qquad \forall\varepsilon \in (0,a^2], t \in [0,T_a]. \tag{7.3}$$

Therefore, the right–hand side of equation (4.1) is uniformly (in ε) bounded in the set $\{(x,t) \in Q_{T_a} : |x| < r_1 + a/4 \text{ or } |x| > r_2 - a/4\}$, so that (7.2) holds in the set $\{(x,t) \in Q_{T_a} : |x| < r_1 + a/8 \text{ or } |x| > r_2 - a/8\}$, by the standard parabolic estimates [15, 20].

It now remains to consider the case when

$$(x,t) \in \Omega_a \times [0,T_a], \qquad \Omega_a := \{x \in \mathbf{R}^N : r_1 + a/8 \le |x| \le r_2 - a/8\}. \tag{7.4}$$

Write u^ε as the sum of u_1, u_2, and u_3 as in (6.4). In view of the estimate (6.7), we need only consider u_3.

Decompose u_3 into the sum of A, B, and C as in (6.13). One can easily conclude that B is smooth since its kernel $\frac{\partial G}{\partial n}(x,t;\xi,\tau)$ is smooth when $\xi \in \partial\Omega$ and $x \in \Omega_a$.

Next we estimate $C(x,t)$. As in the previous section, we can write C as

$$C(x,t) = I + \cdots$$

where

$$I := \int_0^t \int_{-\frac{a}{4\varepsilon}}^{\frac{a}{4\varepsilon}} \Psi'(\eta)\Big[\widetilde{G}_{r'}(r,t;R^\varepsilon(\varepsilon\eta,t),\tau) - \widetilde{G}_{r'}(r,t;R^\varepsilon(\varepsilon\eta,\tau),\tau)\Big]d\eta d\tau$$

and $\cdots$ are smooth terms since their integrands are smooth if $x \in \Omega_a$.

To estimate I, write I as

$$I = \int_0^t d\tau \int_{-\frac{a}{4\varepsilon}}^{\frac{a}{4\varepsilon}} \Psi'(\eta)d\eta \int_{R^\varepsilon(\varepsilon\eta,\tau)}^{R^\varepsilon(\varepsilon\eta,t)} \widetilde{G}_{r'r'}(r,t;r',\tau)dr'.$$

Then, for every $x_1, x_2 \in \Omega_a$, one has the estimate

$$\begin{aligned}
&|I(x_1,t) - I(x_2,t)| \\
\le & \int_0^t d\tau \int_{-\frac{a}{4\varepsilon}}^{\frac{a}{4\varepsilon}} \Psi'(\eta)d\eta \Big| \int_{R^\varepsilon(\varepsilon\eta,\tau)}^{R^\varepsilon(\varepsilon\eta,t)} \|\widetilde{G}_{r'r'}(\cdot,t,r',\tau)\|_{C^\alpha(\Omega_a)}|x_1 - x_2|^\alpha dr'\Big| \\
\le & \int_0^t d\tau \int_{-\frac{a}{4\varepsilon}}^{\frac{a}{4\varepsilon}} \Psi'(\eta)d\eta |R^\varepsilon(\varepsilon\eta,t) - R^\varepsilon(\varepsilon\eta,\tau)| C_a(t-\tau)^{-\frac{3}{2}-\frac{\alpha}{2}}|x_1 - x_2|^\alpha \\
\le & \int_0^t d\tau \int_{-\frac{a}{4\varepsilon}}^{\frac{a}{4\varepsilon}} \Psi'(\eta)d\eta C_\beta |t-\tau|^\beta (t-\tau)^{-\frac{3}{2}-\frac{\alpha}{2}}|x_1 - x_2|^\alpha \quad \text{(by Lemma 5.3)} \\
\le & C_{\beta,a} t^{\beta - \frac{1+\alpha}{2}} |x_1 - x_2|^\alpha
\end{aligned}$$

for all $\beta \in (\frac{1+\alpha}{2}, 1)$. Similarly, one can show that

$$|I(x,t_1) - I(x,t_2)| \le C_{\alpha,a}|t_1 - t_2|^{\alpha/2}, \qquad \forall x \in \Omega_a,\, 0 \le t_1 \le t_2 \le T_a,$$

so that

$$\|I\|_{C^{\alpha,\alpha/2}(Q_t^a)} \le C_{\alpha,a}.$$

Therefore, the function $C(x,t)$ is uniformly (in ε) Hölder continuous.

Finally, we estimate A. By writing it as

$$\begin{aligned}
A =& \int_{r_1}^{r_2} \widetilde{G}(r,t;r',0)\Big[\Psi\Big(\frac{Z^\varepsilon(r',t)}{\varepsilon}\Big) - \Psi\Big(\frac{Z^\varepsilon(r',0)}{\varepsilon}\Big)\Big]dr' \\
=& \int_{Z^\varepsilon(r_1,t)/\varepsilon}^{Z^\varepsilon(r_2,t)/\varepsilon} \widetilde{G}(r,t;R^\varepsilon(\varepsilon\eta,t),0)\Psi(\eta)\frac{\varepsilon}{Z^\varepsilon{}_r(R^\varepsilon(\varepsilon\eta,t),t)}d\eta \\
&- \int_{Z^\varepsilon(r_1,0)/\varepsilon}^{Z^\varepsilon(r_2,0)/\varepsilon} \widetilde{G}(r,t;R^\varepsilon(\varepsilon\eta,0),0)\Psi(\eta)\varepsilon d\eta,
\end{aligned}$$

One can use the same method as in estimating $C(x,t)$ to conclude that

$$\|A\|_{C^{\alpha,\alpha/2}(\Omega_a\times[0,T_a]} \le C_{\alpha,a}.$$

This completes the proof of Theorem 7.1.

8. Convergence to the sharp interface

In this section, we shall complete the proof of Theorem 4.1; i.e., we shall show that u^ε tends to the solution of the sharp interface problem as $\varepsilon \searrow 0$, as long as the interface of the solution of the sharp interface problem does not touch the external walls.

By the estimates obtained in Theorem 6.1, Theorem 6.2, and Theorem 7.1, there exists, for every sequence $\{\varepsilon_j\}_{j=1}^\infty$ satisfying $\varepsilon_j \searrow 0$ as $j \to \infty$, a subsequence, which, for simplicity, we still denote by $\{\varepsilon_j\}_{j=1}^\infty$, such that for all $\alpha \in (0,1)$ and $a > 0$ sufficient small,

$$u^{\varepsilon_j}(x,t) \longrightarrow u(x,t) \qquad \text{uniformly in } C^{\alpha,\alpha/2}(Q_{T_a}), \tag{8.1}$$

$$Z^{\varepsilon_j}(x,t) \longrightarrow Z(x,t) \qquad \text{uniformly in } C^{1+\alpha,(1+\alpha)/2}(Q_T), \tag{8.2}$$

$$R^{\varepsilon_j}(0,t) \longrightarrow S(t) \qquad \text{uniformly in } C^{(1+\alpha)/2}([0,T]) \tag{8.3}$$

for some $u \in C^{\alpha,\alpha/2}(Q_{T_a})$, $Z \in C^{1+\alpha,(1+\alpha)/2}(Q_T)$, and $S \in C^{(1+\alpha)/2}([0,T])$, where T_a is defined in (7.1).

In the following, we shall assume that $a > 0$ is a fixed small constant.

Note that the interface for the solution $(u^\varepsilon,\ \varphi^\varepsilon)$ of (4.1)–(4.5) is given by

$$\Gamma^\varepsilon := \{(x,t) \in Q_T : Z^\varepsilon(x,t) = 0\} = \{(x,t) \in Q_T : |x| = R^\varepsilon(0,t)\},$$

so that (8.3) indicates the interface Γ^{ε_j} convergences to $\Gamma := \{(x,t) \in Q_T : |x| = S(t)\}$ as $\varepsilon_j \searrow 0$.

To prove the main theorem (Theorem 4.1), we need show that (u,Γ) is the unique solution to the sharp interface problem (4.15)–(4.17). This will be done in Theorem 8.1 and Theorem 8.2 below.

LEMMA 8.1. *Let Z, S be as in (8.2), (8.3). Then*

$$Z(x,t) = |x| - S(t), \qquad (x,t) \in Q_T, \tag{8.4}$$

$$\lim_{j\to\infty} \varphi^{\varepsilon_j}(x,t) = \begin{cases} 1 & \text{if } |x| > R(t),\ t \in [0,T], \\ -1 & \text{if } |x| < R(t),\ t \in [0,T]. \end{cases} \tag{8.5}$$

The assertion (8.4) is a consequence of (5.14) and the uniform convergence of R^ε in (8.3) whereas (8.5) follows from the representation $\varphi^\varepsilon = \Psi(Z^\varepsilon/\varepsilon)$, the uniform convergence of Z^ε in (8.2), and equation (8.4).

The following theorem concerning the motion of the interface is a key feature of the equation (4.1), the Cahn–Allen equation [2, 5, 13, 14, 16, 23–26].

THEOREM 8.1. *The function S is of $C^{1+\alpha}([0,T_a])$ and satisfies*

$$\dot{S}(t) = -\frac{N-1}{S(t)} - \beta u(S(t),t), \qquad t \in [0,T_a], \tag{8.6}$$

$$S(0) = r_0 \tag{8.7}$$

where β is a constant defined in (8.19) below.

Proof. We need only show (8.6) since (8.7) follows from the equation $Z^\varepsilon(x,0) = r - r_0$.

In case u is Lipschitz in x (therefore the solution of (8.6), (8.7) is unique), one can directly use the method developed by Chen in [13, 14] to prove the theorem. Since up until now we only have Hölder estimate for u^ε, we need some modifications to the method developed in [14].

The idea of the proof is to construct, for any $t_0 \in [0, T_a)$ and $\delta > 0$, a supersolution $\varphi^{\varepsilon,\delta,t_0}$ to (4.1) in

$$Q_{t_0,T_a} := \Omega \times [t_0, T_a],$$

where the interface of $\varphi^{\varepsilon,\delta,t_0}$ (the zero level set of $\varphi^{\varepsilon,\delta,t_0}$) is located at $|x| = S^{\varepsilon,\delta,t_0}(t)$ and $S^{\varepsilon,\delta,t_0}(t)$ is a solution to the ODE

$$\frac{d}{dt} S^{\varepsilon,\delta,t_0}(t) = -\frac{N-1}{S^{\varepsilon,\delta,t_0}(t)} - \beta \widetilde{u^\varepsilon}(S^{\varepsilon,\delta,t_0}(t), t) - \delta, \qquad t \in [t_0, T], \tag{8.8}$$

$$S^{\varepsilon,\delta,t_0}(t_0) = S(t_0) - 2\delta \tag{8.9}$$

where $\widetilde{u^\varepsilon}$ is a mollifier of u^ε defined in (8.10) below. By first letting $\varepsilon \to 0$ and then $\delta \to 0$, we can conclude that S is a supersolution of (8.6). After using a similar argument to conclude that S is a subsolution of (8.6), one obtains (8.6).

In the following, we shall identify ε_j with ε.

We start by modifying the function u^ε. Let $\hat{u}^\varepsilon$ be a radially symmetric $C^{3/4,3/8}$ extension of u^ε in $\mathbf{R}^N \times \mathbf{R}^1$ and $\zeta(x,t)$ be a nonnegative smooth function supported in the unit ball and of unit mass. We define the mollifier $\widetilde{u^\varepsilon}$ of u^ε by

$$\widetilde{u^\varepsilon}(x,t) := \frac{1}{\varepsilon^{\frac{N+2}{2}}} \int_{\mathbf{R}^N} dy \int_{\mathbf{R}^1} d\tau\, \zeta\Big(\frac{y-x}{\varepsilon^{1/2}}, \frac{\tau - t}{\varepsilon}\Big) \hat{u}^\varepsilon(y,\tau). \tag{8.10}$$

One can show directly, by using the $C^{3/4,3/8}$ estimate for u^ε (Theorem 7.1), that $\widetilde{u^\varepsilon}$ is radially symmetric and satisfies

$$|\widetilde{u^\varepsilon}(x,t) - u^\varepsilon(x,t)| \le \sup_{|y-x|+|t-\tau|^{1/2} \le \varepsilon^{1/2}} |\hat{u}^\varepsilon(y,\tau) - \hat{u}^\varepsilon(x,t)| \le C\varepsilon^{3/8}, \tag{8.11}$$

$$\|\varepsilon^{1/2}\widetilde{u^\varepsilon}_r,\ \varepsilon\widetilde{u^\varepsilon}_{rr},\ \varepsilon\widetilde{u^\varepsilon}_t\|_{C^0(Q_{T_a})} \le C, \tag{8.12}$$

$$\|\widetilde{u^\varepsilon}\|_{C^{3/4,3/8}(Q_{T_a})} \le C \tag{8.13}$$

for some constant C independent of ε.

Next, we define the constant β which appeared in (8.8). Set

$$F(\varphi, \lambda, \mu) := (1 - \varphi^2)(2\varphi + \lambda(1 - \varphi^2)) + \mu. \tag{8.14}$$

Then, there exists a constant $\mu_0 > 0$ such that for every $\lambda \in [-1,1]$ and $\mu \in [0,\mu_0]$ the algebraic equation, for φ,

$$F(\varphi,\lambda,\mu) = 0, \tag{8.15}$$

has exactly three solutions: $h^-(\lambda,\mu)$, $h^0(\lambda,\mu)$, and $h^+(\lambda,\mu)$, and they satisfy

$$\begin{aligned} &h^-(\lambda,\mu) < h^0(\lambda,\mu) < h^+(\lambda,\mu), \\ &h^-(\lambda,\mu) < 0 < h^+(\lambda,\mu), \\ &h^\pm(\lambda,\mu) \geq \pm 1 + b\mu \end{aligned} \tag{8.16}$$

for some constant $b > 0$. By a result of Aronson and Weinberger [3], there exists a unique solution $(\Lambda(\lambda,\mu), Q(\lambda,\mu,\rho))$ to the nonlinear eigenvalue problem

$$Q_{\rho\rho} - \Lambda Q_\rho + F(Q,\lambda,\mu) = 0, \tag{8.17}$$

$$Q(\lambda,\mu,\pm\infty) = h^\pm(\lambda,\mu), \qquad Q(\lambda,\mu,0) = 0 \tag{8.18}$$

for any $\lambda \in [-1,1]$ and $\mu \in [0,\mu_0]$. We define the constant β by

$$\beta := \frac{\partial\Lambda}{\partial\lambda}(0,0). \tag{8.19}$$

Some properties of the solution $(\Lambda(\lambda,\mu), Q(\lambda,\mu,\rho))$ are stated in the following lemma which has been proven in the appendix of [18]

LEMMA 8.2. *There exist positive constants c and A such that for any $\lambda \in [-1,1]$ and $\mu \in [0,\mu_0]$, the solution (Λ, Q) to (8.17), (8.18) satisfies*

(8.20)
$$Q_\rho > 0, \qquad \forall \rho \in R^1,$$

(8.21)
$$\sup_{\rho\in\mathbf{R}^1} |Q_\rho,\ \rho Q_\rho,\ Q_\lambda,\ Q_{\lambda\rho},\ Q_{\lambda\lambda},\ \Lambda_\lambda,\ \Lambda_\mu,\ \Lambda_{\lambda\lambda},\ \Lambda_{\lambda\mu},\ \Lambda_{\mu\mu}| \leq A,$$

(8.22)
$$Q(\lambda,\mu,\rho) \geq h^+(\lambda,\mu) - Ae^{-c\rho}, \qquad \forall \rho > 0.$$

Since the right–hand side of (8.8) is smooth (as a function of $S^{\varepsilon,\delta,t_0}$ and t), the ODE system (8.8), (8.9) has a unique (local) solution $S^{\varepsilon,\delta,t_0}$ and the solution exists as long as it remains in the interval (r_1, r_2). According to the definition of T_a in (7.1), we can assume, without loss of generality, that $S^{\varepsilon,\delta,t_0}$ exists in $[0, T_a]$ and

$$r_1 - a/2 \leq S^{\varepsilon,\delta,t_0}(t) < r_2, \qquad \forall t \in [0, T_a]. \tag{8.23}$$

We now define $\varphi^{\varepsilon,\delta,t_0}$ by

$$\varphi^{\varepsilon,\delta,t_0} := Q\Big(\varepsilon\widetilde{u^\varepsilon}(x,t), \varepsilon^{9/8}, \frac{|x| - S^{\varepsilon,\delta,t_0}(t)}{\varepsilon}\Big), \qquad (x,t) \in Q_{t_0,T_a}. \tag{8.24}$$

Since for small enough ε, we have

$$|\varepsilon\widetilde{u^\varepsilon}| \leq 1 \qquad \text{and} \qquad \varepsilon^{9/8} \leq \mu_0,$$

the first two arguments for the function Q in (8.24) are in the range of its definition, so that $\varphi^{\varepsilon,\delta,t_0}$ is well–defined.

To complete the proof of Theorem 8.1, we need the following Lemma.

LEMMA 8.3. *For every $\delta > 0$, there exists a constant $\varepsilon_\delta > 0$ such that $\forall \varepsilon \in (0, \varepsilon_\delta]$, the function $\varphi^{\varepsilon,\delta,t_0}$ defined in (8.24) satisfies*

$$\varphi^{\varepsilon,\delta,t_0}(x,t) \geq \varphi^{\varepsilon}(x,t), \qquad (x,t) \in Q_{t_0,T}. \tag{8.25}$$

Consequently,

$$S^{\varepsilon,\delta,t_0}(t) \leq R^{\varepsilon}(0,t), \qquad t \in [t_0, T]. \tag{8.26}$$

We continue the proof of theorem of 8.1. Since the Hölder norm of the right–hand side of (8.8) is bounded independent of ε, the $C^{1+\alpha}([t_0, T_a])$ norm of $S^{\varepsilon,\delta,t_0}$ is also bounded, so that the set $\{S^{\varepsilon,\delta,t_0}\}_{0<\delta\leq\delta_0, 0<\varepsilon\leq\varepsilon_\delta}$ is equicontinuous in $C^1[t_0, T_a]$. Therefore,

$$S_{t_0}(t) := \overline{\lim_{\delta\to 0+}}\ \overline{\lim_{\varepsilon\to 0+}}\ S^{\varepsilon,\delta,t_0}(t)$$

exists, and S_{t_0} is in $C^{1+\alpha}[t_0, T_a]$ and satisfies

$$\dot{S}_{t_0}(t_0) = -\frac{N-1}{S(t_0)} - \beta u(S(t_0), t_0).$$

Hence,

$$\begin{aligned}
\dot{S}^-(t_0) &:= \underline{\lim}_{h\to 0+} \frac{S(t_0+h) - S(t_0)}{h} \\
&= \underline{\lim}_{h\to 0+} \lim_{\varepsilon\to 0+} \frac{R^{\varepsilon}(0, t_0+h) - S(t_0)}{h} && \text{(by (8.3))} \\
&\geq \underline{\lim}_{h\to 0+} \overline{\lim_{\delta\to 0+}}\ \overline{\lim_{\varepsilon\to 0+}} \frac{S^{\varepsilon,\delta,t_0}(t_0+h) - S(t_0)}{h} && \text{(by (8.26))} \\
&= \dot{S}_{t_0}(t_0) = -\frac{N-1}{S(t_0)} - \beta u(S(t_0), t_0).
\end{aligned} \tag{8.27}$$

That is, S is a supersolution of (8.6). Similarly, we can show S is a subsolution of (8.6), and therefore, S is a solution of (8.6).

To complete the proof of Theorem 8.1, it remains to prove Lemma 8.3. To do this, we need an auxiliary lemma.

LEMMA 8.4. *There exists a positive constant $\varepsilon_\delta > 0$ such that for all $\varepsilon \in (0, \varepsilon_\delta]$, one has*

$$\delta\varepsilon + \varepsilon b\widetilde{u^{\varepsilon}} - \Lambda(\varepsilon\widetilde{u^{\varepsilon}}, \varepsilon^{9/8}) \geq 0, \qquad (x,t) \in Q_T. \tag{8.28}$$

Proof. By Taylor's expansion, one has

$$\begin{aligned}
\Lambda(\varepsilon\widetilde{u^{\varepsilon}}, \varepsilon^{9/8}) &= \Lambda(0,0) + \frac{\partial\Lambda}{\partial\lambda}(0,0)\varepsilon\widetilde{u^{\varepsilon}} + \frac{\partial\Lambda}{\partial\mu}(0,0)\varepsilon^{9/8} + O(|\varepsilon\widetilde{u^{\varepsilon}}|^2 + |\varepsilon^{9/8}|^2) \\
&= \varepsilon\beta\widetilde{u^{\varepsilon}} + O(\varepsilon^{9/8})
\end{aligned}$$

since $\Lambda(0,0)=0$ and $\widetilde{u^\varepsilon}$ is bounded. The inequality (8.28) thus holds for ε sufficient small.

Proof of Lemma 8.3. We need only prove (8.25) since (8.26) follows from the fact that $R^\varepsilon(0,t)$ and $S^{\varepsilon,\delta,t_0}(t)$ are the zero level sets of φ^ε and $\varphi^{\varepsilon,\delta,t_0}$ respectively.

By means of a comparison principle for semilinear parabolic equations, one can prove (8.25) provided that one can show the following:

(8.29) $\varphi^\varepsilon(x,t_0) \le \varphi^{\varepsilon,\delta,t_0}(x,t_0), \qquad x\in\Omega,$

(8.30) $\varphi^\varepsilon(x,t) \le \varphi^{\varepsilon,\delta,t_0}(x,t), \qquad (x,t)\in\partial\Omega\times[t_0,T_a],$

(8.31) $\mathcal{L}\varphi^{\varepsilon,\delta,t_0} := \varphi_t^{\varepsilon,\delta,t_0} - \Delta\varphi^{\varepsilon,\delta,t_0} - \frac{1}{\varepsilon^2}F(\varphi^{\varepsilon,\delta,t_0},\varepsilon u^\varepsilon,0)\ge 0, \quad (x,t)\in Q_{t_0,T_a}.$

To prove (8.29), consider two cases:

(i) $|x| < S(t_0)-\delta$;

(ii) $|x| \ge S(t_0)-\delta$.

In case (i), one has the bound

$$Z^\varepsilon(x,t) = Z^\varepsilon(x,t) - Z^\varepsilon(R^\varepsilon(0,t),t) = Z^\varepsilon{}_r(\xi,t)(|x|-R^\varepsilon(0,t)) \le -\frac{\delta}{4}$$

by the mean value theorem, the estimate $1/2 < Z^\varepsilon{}_r < 2$ (Lemma 5.1) and the fact

$$|R^\varepsilon(0,t) - S(t)| \le \delta/2 \tag{8.32}$$

if ε (actually ε_j) is small enough. Therefore, one has

$$\begin{aligned}
\varphi^\varepsilon(x,t_0) &= \Psi\Big(\frac{Z^\varepsilon(x,t_0)}{\varepsilon}\Big)\\
&\le \Psi\Big(-\frac{\delta}{4\varepsilon}\Big) \le -1+2e^{-\frac{\delta}{4\varepsilon}}\\
&\le -1+b\varepsilon^{9/8} \le h^-(\widetilde{\varepsilon u^\varepsilon},\varepsilon^{9/8}) \quad \text{(by (8.16))}\\
&\le \varphi^{\varepsilon,\delta,t_0}(x,t_0).
\end{aligned}\tag{8.33}$$

In case (ii), we can use the initial condition for $S^{\varepsilon,\delta,t_0}$ in (8.8) to conclude that

$$\begin{aligned}
\varphi^{\varepsilon,\delta,t_0}(x,t_0) &= Q\Big(\frac{|x|-S(t_0)+2\delta}{\varepsilon},\widetilde{\varepsilon u^\varepsilon},\varepsilon^{9/8}\Big)\\
&\ge Q\Big(\frac{\delta}{\varepsilon},\widetilde{\varepsilon u^\varepsilon},\varepsilon^{9/8}\Big)\\
&\ge h^+(\widetilde{\varepsilon u^\varepsilon},\varepsilon^{9/8}) - Ae^{-c\delta/\varepsilon}\\
&\ge 1+b\varepsilon^{9/8} - Ae^{-c\delta/\varepsilon}\\
&\ge 1 \ge \varphi^\varepsilon(x,t_0)
\end{aligned}\tag{8.34}$$

where (8.22) and (8.16) have been used. Combining (8.33) with (8.34), inequality (8.29) follows.

Similarly, we can show that (8.30) holds by using (8.23) and the definition of T_a.

Finally we verify (8.31). We compute the identity

$$\begin{aligned}\mathcal{L}\varphi^{\varepsilon,\delta,t_0} &= -\frac{1}{\varepsilon}Q_\rho S^{\varepsilon,\delta,t_0}{}_t + \varepsilon Q_\lambda \widetilde{u^\varepsilon}_t \\ &- \Big[\frac{1}{\varepsilon^2}Q_{\rho\rho} + \frac{1}{\varepsilon}\frac{N-1}{|x|}Q_\rho + 2Q_{\rho\lambda}\widetilde{u^\varepsilon}_r + \varepsilon Q_\lambda \Delta\widetilde{u^\varepsilon} + \varepsilon^2 Q_{\lambda\lambda}\widetilde{u^\varepsilon}_r^2\Big] \\ &- \frac{1}{\varepsilon^2}\Big[F(Q,\varepsilon u^\varepsilon,0) + F(Q,\varepsilon\widetilde{u^\varepsilon},\varepsilon^{9/8}) - F(Q,\varepsilon\widetilde{u^\varepsilon}^\varepsilon,\varepsilon^{9/8})\Big].\end{aligned}$$

Using the equation for Q [(8.17)] and the definition of F [(8.14)], one obtains

$$\begin{aligned}\mathcal{L}\varphi^{\varepsilon,\delta,t_0} =& \frac{Q_\rho}{\varepsilon}\Big[-\frac{d}{dt}S^{\varepsilon,\delta,t_0} - \frac{\Lambda(\varepsilon\widetilde{u^\varepsilon},\varepsilon^{9/8})}{\varepsilon} - \frac{N-1}{|x|}\Big] \\ &- \Big[Q_\lambda(\varepsilon\Delta\widetilde{u^\varepsilon} - \varepsilon\widetilde{u^\varepsilon}_t) + 2Q_{\rho\lambda}\widetilde{u^\varepsilon}_r + \varepsilon^2 Q_{\lambda\lambda}\widetilde{u^\varepsilon}_r^2\Big] \\ &- \frac{1}{\varepsilon^2}\Big[\varepsilon(u^\varepsilon - \widetilde{u^\varepsilon})(1-Q^2)^2 - \varepsilon^{9/8}\Big] \\ :=& I + II + III.\end{aligned} \tag{8.35}$$

One can estimate II and III as

$$II \geq -C\varepsilon^{-1/2}, \tag{8.36}$$

$$III \geq -C\varepsilon^{-1/2} + \varepsilon^{-7/8} \tag{8.37}$$

by using (8.11), (8.12) and (8.21).

Since $S^{\varepsilon,\delta,t_0}$ satisfies the ODE (8.8), we can write I as

$$\begin{aligned}I &= \frac{Q_\rho}{\varepsilon}\Big[\frac{N-1}{S^{\varepsilon,\delta,t_0}(t)} + \beta\widetilde{u^\varepsilon}(S^{\varepsilon,\delta,t_0}(t),t) + \delta - \frac{\Lambda(\varepsilon\widetilde{u^\varepsilon},\varepsilon^{9/8})}{\varepsilon} - \frac{N-1}{|x|}\Big] \\ &= \frac{Q_\rho}{\varepsilon}\Big[\frac{\delta\varepsilon + \varepsilon\beta\widetilde{u^\varepsilon} - \Lambda(\varepsilon\widetilde{u^\varepsilon},\varepsilon^{9/8})}{\varepsilon} + \beta\big(\widetilde{u^\varepsilon}(S^{\varepsilon,\delta,t_0}(t),t) - \widetilde{u^\varepsilon}(x,t)\big) \\ &\qquad + (N-1)\frac{|x| - S^{\varepsilon,\delta,t_0}(t)}{|x|S^{\varepsilon,\delta,t_0}(t)}\Big] \\ &\geq -\frac{Q_\rho}{\varepsilon}\Big[0 + C\beta\big|S^{\varepsilon,\delta,t_0}(t) - |x|\big|^{3/4} + C(N-1)\big||x| - S^{\varepsilon,\delta,t_0}(t)\big|\Big] \\ &\geq -C\varepsilon^{-1/4}\sup_{\rho\in\mathbf{R}^1}|\rho^{3/4}Q_\rho| \geq -C\varepsilon^{-1/4},\end{aligned} \tag{8.38}$$

where Lemma 8.4, (8.20), (8.13), and (8.21) have been used.

Substituting (8.36)–(8.38) into (8.35), one obtains

$$\mathcal{L}\varphi^{\varepsilon,\delta,t_0} \geq -C\varepsilon^{-1/4} - C\varepsilon^{-1/2} - C\varepsilon^{-1/2} + \varepsilon^{-7/8} \geq 0$$

if ε is sufficient small. This proves (8.31) and completes the proof of Lemma 8.2 and also Theorem 8.1. □

THEOREM 8.2. *Let*

$$\begin{aligned}Q_1 :=&\{(x,t) \in Q_{T_a} : |x| < S(t)\},\\ Q_2 :=&\{(x,t) \in Q_{T_a} : |x| > S(t)\},\\ \Gamma :=&\{(x,t) \in Q_{T_a} : |x| = S(t)\}.\end{aligned}$$

Then, the function u defined in (8.1) satisfies

$$u \in C^{1+\alpha,(1+\alpha)/2}(\overline{Q_1}) \cup C^{1+\alpha,(1+\alpha)/2}(\overline{Q_2}), \tag{8.39}$$

$$u_t - \Delta u = 0 \quad \text{in } Q_1 \cup Q_2, \tag{8.40}$$

$$\dot{S}(t) = \big[u_r\big]_+^- \quad \text{on } \Gamma. \tag{8.41}$$

Proof. Let $\Gamma(x,t;\xi,\tau)$ be the fundamental solution of the heat operator $\partial_t - \Delta$, and $w(x,t)$ be the function defined by the surface potential

$$w(x,t) := \int_0^t d\tau \int_{|\xi|=S(\tau)} \dot{S}(\tau)\Gamma(x,t;\xi,\tau)dS_\xi.$$

Then, since $\Gamma \in C^{1+\alpha}$, we know [20, Chapter 5] that

$$w \in C^{1+\alpha,(1+\alpha)/2}(\overline{Q_1}) \cup C^{1+\alpha,(1+\alpha)/2}(\overline{Q_2}), \tag{8.42}$$

$$w_t - \Delta w = 0 \quad \text{in } Q_1 \cup Q_2, \tag{8.43}$$

$$\big[w_r\big]_+^- = \dot{S}(t) \quad \text{on } \Gamma. \tag{8.44}$$

Therefore, for every $\zeta(x,t) \in C_0^\infty(Q_{T_a})$, one has

$$\begin{aligned}0 &= \iint_{Q_{T_a}} \big[u_t^{\varepsilon_j} - \Delta u^{\varepsilon_j} + \frac{1}{2}\varphi_t^{\varepsilon_j}\big]\zeta - \iint_{Q_1 \cup Q_2} (w_t - \Delta w)\zeta\\ &= \iint_{Q_{T_a}} (u^{\varepsilon_j} - w)(-\zeta_t - \Delta\zeta) - \frac{1}{2}\iint_{Q_{T_a}} \varphi^{\varepsilon_j}\zeta_t + \int_0^{T_a} \dot{S}(\tau)\zeta(S(\tau),\tau)d\tau\end{aligned}$$

Letting $\varepsilon_j \to 0$ and using (8.1) and Lemma 8.1, one gets

$$\begin{aligned}0 &= \iint_{Q_T} (u - w)(-\zeta_t - \Delta\zeta) - \frac{1}{2}\iint_{Q_2} \zeta_t + \frac{1}{2}\iint_{Q_1} \zeta_t + \int_0^T \dot{S}(\tau)\zeta(S(\tau),\tau)d\tau\\ &= \iint_{Q_T} (u - w)(-\zeta_t - \Delta\zeta),\end{aligned}$$

Hence,

$$(u - w)_t - \Delta(u - w) = 0, \qquad \forall (x,t) \in Q_{T_a}, \tag{8.45}$$

and therefore

$$u - w \in C^\infty(Q_{T_a}).$$

Theorem 8.2 thus follows from (8.42)–(8.44). □

Recall that the solution of (4.14)–(4.16) is unique [15], so that (8.1)–(8.3) are valid for all sequence $\varepsilon \to 0$. Letting $a \to 0$, one can conclude, from the definition of T_a, that $T_a \to T^*$, thereby proving Theorem 4.1.

An immediate corollary of the proof of Theorem 4.1 is an analogous result for motion by curvature

$$\dot{S}(t) = -\frac{N-1}{S(t)} \tag{8.46}$$

If we consider the original heat equation (1.9) without scaling out the latent heat, ℓ, i.e.

$$u_t + \frac{\ell}{2}\varphi_t = \tilde{K}\Delta u \tag{8.47}$$

coupled with (1.8), then we can consider the limit as ℓ approaches zero along with ε; in particular

$$|\ell| \leq C\varepsilon. \tag{8.48}$$

With $w \equiv \ell^{-1}u$, it is clear that (w, φ) satisfies (1.8), (1.9) with an additional factor of ε in the last term of (1.8). The uniform bounds on w are then obtained in the same way and (4.17) is attained without the βu term because of the additional factor of ε, to yield the following result.

Corollary 8.3. Let $(u^\varepsilon, \varphi^\varepsilon)$ be the solution of (4.1), (8.47) subject to initial and boundary conditions (4.3)-(4.5). Let $u_0 \in C^{2,1}(Q_T)$ be radially symmetric with $(u_0(x,t)| \leq C_\varepsilon$ and ℓ subject to (8.48). Then the conclusions (4.12) - (4.14) remain valid and $S(t)$ satisfies (8.46).

Remark. In view of the scaling discussed above, the theorem may be stated with the hypothesis

$$|\gamma| \leq C\varepsilon \tag{8.49}$$

replacing the scaling assumptions on u_0 and ℓ.

REFERENCES

[1] N. Alikakos and P. Bates, *On the singular limit in a phase field model of a phase transition*, Ann. Inst. H. Poincarè, 5 (1988), pp. 1–38.

[2] S. Allen and J. Cahn, *A microscopic theory for antiphase boundary motion and its application to antiphase domain coarsening*, Acta Metall, 27 (1979), pp. 1084–1095.

[3] D. G. Aronson and H. F. Weinberger, *Nonlinear diffusion in population genetics, combustion, and nerve propagation, in Partial Differential Equation and Related Topics, ed J. A. Goldstein. Lecture notes in Mathematics*, Springer, New York, 1975, pp. 5–49.

[4] M. S. Berger and L. E. Frankel, *On the asymptotic solution of nonlinear Dirichlet problem*, J. Math. Mech., 19 (1970), pp. 553–585.

[5] L. Bronsard and R. V. Kohn, *Motion by mean curvature as the singular limit of Ginzburg–Landau dynamics*, to appear in J. Diff. Eqns..

[6] G. CAGINALP, *Mathematical models of phase boundaries, Material Instabilities in Continuum Problems and Related Mathematical Problems*, Proc. of 1985–86 Heroit–Watt Conf. ed. J. Ball, Oxford Science, 1988, p. 35-52.

[7] G. CAGINALP, *Stefan and Hele–Shaw type models as asymptotic limits of phase field equations*, Physics Review A, 39 (1989), pp. 887–896.

[8] G. CAGINALP, *An analysis of a phase field model of a free boundary*, Archive for Rational Mechanics and Analysis, 92 (1986), pp. 205–245.

[9] G. CAGINALP AND P. FIFE, *Elliptic problems involving phase boundaries satisfying a curvature condition*, IMA J. of Appl. Math., 38 (1987), pp. 195–217.

[10] G. CAGINALP AND B. MCLEOD, *The interior transition layer for an ordinary differential equation arising from solidification theory*, Quarterly of Appl. Math., 44 (1986), p. 155–168.

[11] G. CAGINALP AND Y. NISHIURA, *The existence of traveling waves for phase field equations and convergence to sharp interface models in singular limit*, Quartly of Appl. Math., 49 (1991), pp. 147–162.

[12] G. CAGINALP AND J. JONES, *A derivation of phase-field model with fluid properties*, (to appear in the Proceedings of the IMA workshop on Evolution of Phase Boundaries, 1991).

[13] XINFU CHEN, *Generation and propagation of interface for reaction–diffusion equations*, to appear in J. Diff. Eqns..

[14] XINFU CHEN, *Generation and propagation of interfaces for reaction–diffusion systems*, IMA preprint #708, University of Minnesota, MN 55455.

[15] XINFU CHEN AND F. REITICH, *Local existence and uniqueness of solutions of the Stefan problem with surface tension and kinetic undercooling*, to appear in J. Math. Anal. and Appl..

[16] XU-YAN CHEN, *Dynamics of interfaces in reaction diffusion systems*, to appear in Hiroshima Math J. Vol. 21, No. 1 , (1991).

[17] S.D. EIDELMAN, *Parabolic Systems*, No. Holland Publ., Amsterdam, 1969.

[18] P. C. FIFE AND L. HSIAO, *The generation and propagation of internal layers*, Nonlinear Anal. TMA, 12 (1988), pp. 19-41.

[19] P. FIFE AND O. PENROSE, *Thermodynamically consistent models of phase–field type for the kinetics of phase transitions*, Phys D, 43 (1990), pp. 44–62.

[20] AVNER FRIEDMAN, *Partial Differential Equations of Parabolic Type*, Prentice–Hall: Englewood Cliffs, NJ, 1964.

[21] S. LUCKHAUS AND L. MODICA, *The Gibbs–Thomson relation within the gradient theory of phase transitions*, Archive for Rational Mechanics and Analysis, 107 (1989), pp. 71–83.

[22] H. MATANO, *Convergence of solutions of one–dimensional semilinear parabolic equations*, J. Math. Kyoto Univ., 18 (1978), pp. 221–227.

[23] P. DEMOTTONI AND M. SCHATZMAN, *Evolution gèometroque d'interfaces*, C. R. A. S., 309 (1989), pp. 453–458.

[24] P. DEMOTTONI AND M. SCHATZMAN, *Development of interfaces in N-dimensional space*, preprint.

[25] J. RUBINSTEIN, P. STERBERG, AND J. B. KELLER, *Fast reaction, slow diffusion and curve shorting*, SIAM., J. Appl. Math., 49 (1989), pp. 116–133.

[26] J. RUBINSTEIN, P. STERBERG, AND J. B. KELLER, *Reaction–diffusion processes and evolution to harmonic maps*, to appear in SIAM, J. Appl. Math..

[27] L. I. RUBINSTEIN, *The Stefan Problem*, AMS Translation, 27, AMS, Providence, 1971.

A PHASE FLUID MODEL: DERIVATION AND NEW INTERFACE RELATION

G. CAGINALP*† AND J. JONES*

Abstract. We develop a very general model of phase boundaries in which the fluid properties play a role. The variables pressure, fluid velocity and specific volume are considered in conjunction with temperature and order parameter using the phase field approach. Upon making some choices and approximations one obtains a system of parabolic differential equations. An asymptotic analysis leads to a new interface relation (generalizing Gibbs-Thomson) which indicates that the front velocity in the kinetic undercooling term should be replaced by the front velocity minus the normal fluid velocity. The temperature term involves a quadratic term due to thermal expansion and isothermal compressibility terms.

1. Introduction. The phase field models encompass a broad spectrum of free boundary problems as their limiting cases [1, 2]. These include the classical Stefan model, the surface tension and kinetic undercooling models (modified Stefan) the Hele-Shaw model for fluid pressure, the Cahn-Allen antiphase boundary model, and motion by mean curvature. Thus, the large diversity of the phenomena and patterns exhibited by these different sharp interface models are contained within a single pair of equations and are obtained as distinguished limits which depend crucially on the physical parameters in the phase field model.

In this paper, we derive and develop a set of equations that extend the basic phase field equations, which comprise the temperature and phase variables, to a system which also includes fluid velocity, pressure and density. We present a brief review of the issues which relate to the temperature and phase (along the lines of [2]) before introducing the fluid variables.

In the mathematical literature, the original attempts to analyze these problems (dating back to Lamé and Clapeyron in 1831 [3]) involved the use of a single variable, namely temperature $T(t, x)$ as a function of time $t \in \mathbf{R}^+$ and space $x \in \Omega \subset \mathbf{R}^N$. In retrospect an important feature of this classical Stefan approach is that it endeavors to treat a sharp phase boundary $\Gamma_s(t)$ by means of conditions imposed on the temperature function <u>without</u> the use of a separate variable for phase. A heat diffusion equation is also satisfied on both sides of the boundary. In particular, the classical Stefan problem is posed as finding suitably regular $T(t, x)$ and $\Gamma_s(t)$ which satisfy the equations

$$T_t = K\Delta T \quad \text{in} \quad \Omega \setminus \Gamma_s(t) \tag{1.1}$$

$$\ell v = -K[\nabla T \cdot \hat{n}]_-^+ \quad \text{on} \quad \Gamma_s(t) \tag{1.2}$$

$$T = 0 \quad \text{on} \quad \Gamma_s(t) \tag{1.3}$$

*Department of Mathematics and Statistics, University of Pittsburgh, Pittsburgh, PA 15260
†Supported by NSF Grant DMS 9002242

where K and ℓ are dimensionless constants related to (heat) diffusivity and latent heat, respectively. Here, $[\nabla T \cdot \hat{n}]_-^+$ denotes the jump in the normal component of the gradient of temperature across the phase boundary, while v denotes the (normal) velocity of the interface with the sign convention that it is positive if the motion is directed toward the liquid (denoted +) and negative if directed toward the solid (denoted −).

This simple distinction of the phases as a function of the sign of the temperature is a physical oversimplification which has serious consequences for the motion and shape of the boundary in many situations. One of the physical effects neglected by the classical Stefan model is that of interfacial tension, which is a crucial factor in the stability properties of the phase boundary. As noted by Gibbs [4] in the last century (in a slightly different context) an immediate consequence of this surface tension, σ, is to modify the temperature at the interface so that

$$T(t,x) = -\frac{\sigma}{[S]_E}\kappa(t,x) \qquad x \in \Gamma_s(t) \tag{1.4}$$

where $[S]_E$ is the difference in equilibrium entropy density per unit mass and κ is the sum of principal curvatures.

The phenomena exhibited by (1.4) is an example of equilibrium "supercooling", in which the liquid occurs below the freezing temperature due to the geometry of the region [5, 6]. In recent decades, materials scientists have also discovered [7] the possibility of "kinetic supercooling" in which the temperature at the interface is lowered due to the velocity, so that (1.4) is modified to the form

$$T(t,x) = -\frac{\sigma}{[S]_E}\{\kappa(t,x) + \alpha v(t,x)\} \qquad x \in \Gamma_s(t) \tag{1.5}$$

where α is a positive constant.

In principle, one can modify the classical Stefan model [(1.1) - (1.3)] by replacing (1.3) with (1.4) or (1.5). Since the phase of each spatial point is no longer determined by the temperature, this poses both a theoretical and practical difficulty which is most directly resolved by tracking the interface. Additional difficulties arise with topological changes in the interface.

We note also that equations (1.1) and (1.2) are equivalent, in a weak sense, to the single equation [8]

$$u_t + \frac{\ell}{2}\varphi_t^{(s)} = K\Delta u \qquad \varphi^{(s)} \equiv \begin{cases} +1 & \text{liquid} \\ -1 & \text{solid.} \end{cases} \tag{1.6}$$

While this discussion has been based on equal diffusivities in the two phases, one can also consider the more general case in which these diffusivities differ. Also, the liquid-solid terminology is used for convenience since the phase transition can be quite general.

Interfacial relations such as (1.4) or (1.5) can be derived from different perspectives. One of these is that the terms on the right hand side of (1.5) arise due to a scaling limit in which the length scale which measures the width of the interface approaches zero (on a macroscopic scale) along with other microscopic parameters. From this perspective, the microscopically nonzero interfacial thickness and the nonzero temperature at the interface are intrinsically linked, with the magnitude of the temperature deviation depending on the scaling relations.

This interpretation demonstrates the necessity for a phase function $\varphi(t,x)$ which has a transition layer behavior instead of the step function form of $\varphi^{(s)}$. Thus, equation (1.6) may be expected to remain valid since it expresses energy conservation, while a second equation is needed for a complete system. Given a suitable free energy $\mathcal{F}(T,\varphi)$, statistical mechanics stipulates that the equilibrium value is determined as a minimizer of $\mathcal{F}$ so that $\mathcal{F}_\varphi = 0$. A basic dynamical ansatz is that the change is directed toward equilibrium with a "force" which is proportional to the extent by which it deviates from equilibrium, i.e.,

$$\tau\varphi_t = -\mathcal{F}_\varphi \tag{1.7}$$

Given a Landau-Ginzburg type free energy (see [9] and references therein)

$$\mathcal{F}\{T,\varphi\} = \int \{\frac{1}{2}\xi^2(\nabla\varphi)^2 + \frac{1}{8a}(\varphi^2-1)^2 - 2u\varphi\}d^N x \tag{1.8}$$

where ξ and a are microscopic parameters which can be related to macroscopic constants, one can then write the system of equations

$$T_t + \frac{\ell}{2}\varphi_t = K\Delta T \tag{1.9}$$

$$\alpha\xi^2\varphi_t = \xi^2\Delta\varphi + \frac{1}{2a}(\varphi-\varphi^3) + 2T. \tag{1.10}$$

This system can then be studied subject to suitable initial and boundary conditions.

It has been shown [1, 2, 9] that the system [(1.9), (1.1)] can be used to derive interfacial relations such as (1.4) or (1.5). In fact, within an appropriate scaling of the parameters in these equations, one can show that the Stefan problem [(1.1) - (1.3)] or surface tension model [(1.), (1.2), (1.4)] emerge as distinguished limits. In addition to the Hele-Shaw and Cahn-Allen limits, the motion by mean curvature limit, which consists of just (1.5) with $T = 0$, i.e.

$$v(t,x) = -\frac{1}{\alpha}\kappa(t,x), \qquad x \in \Gamma_s(t), \tag{1.11}$$

can be attained by also allowing ℓ to vanish.

In this paper, we begin a study of phase boundaries in which fluid properties also play a role. In addition to a physically consistent derivation of a set of equations, our goal is to determine a new interface relation, generalizing (1.5), which relates the fluid variables and equilibrium thermodynamic constants to the temperature at the interface. The outline of the paper is as follows. In Section 2 we formulate a very general model which incorporates the physical concepts. In Section 3 we obtain a specific model upon making some thermodynamic simplifications and choices.

In Section 4, we utilize matched asymptotic analysis to deduce a new interface relation [namely (4.2)]. An interesting feature of this relation is that the normal velocity in (1.5) must be replaced by the difference between the interface and fluid velocities. The contribution due to the entropy difference between phases now involves a term involving the square of the temperature, thereby introducing a new type of nonlinearity into this equation.

In Section 5, we study the system of equations derived in Section 3 for its consequences for plane waves, and obtain some simple relations between the variables. In particular, we find that the system of five equations reduces to just two (phase and any one of the other variables) in this case.

2. The general model. We first consider a very general system of equations before simplifying some aspects in order to facilitate analysis. Throughout this discussion we assume an equilibrium phase diagram (e.g. pressure-volume) so that our reference point will be any point on the coexistence curve between the two phases (denoted by the subscript E). We let $T^{(a)}$ and $P^{(a)}$ be the (absolute) temperature and pressure while $T \equiv T^{(a)} - T_E$ and $P \equiv P^{(a)} - P_E$ denote the temperature and pressure deviations from these equilibrium values. The other variables are the specific volume, $w(\equiv 1/\text{density})$, the fluid velocity $\vec{u}$, and the order parameter, λ. In the general case, we also define a stress tensor, $\vec{\vec{P}}$, which is coupled to the other variables. The fluid velocity, $\vec{u}(\vec{x}, t)$ is the velocity at a given time t and spatial point $\vec{x}$. That is, $\vec{x}$ refers to fixed points in space and not to fixed "particles" of fluid. The standard notation $d/dt \equiv \partial/\partial t + \vec{u} \cdot \vec{\nabla}$ for the convective derivative is used below.

The dynamic equations governing such a system must consist of conservation of mass, momentum, energy, an equation of state and a kinetic equation for the order parameter.

Thus, one has the continuity equation

$$\frac{dw}{dt} = w\vec{\nabla} \cdot \vec{u} \tag{2.1}$$

and the general form of Newton's law (neglecting gravitational forces)

$$\frac{d\vec{u}}{dt} = -w\vec{\nabla} \cdot \vec{\vec{P}}, \qquad \text{or,} \qquad \frac{du_j}{dt} = -w\sum_i \frac{\partial}{\partial x_i}\sigma_{ji} \tag{2.2}$$

The heat balance equation is given by

$$(T + T_E)\frac{dS}{dt} = w\vec{\nabla} \cdot \vec{Q} \tag{2.3}$$

where S is entropy density (per unit mass) and $\vec{Q}$ is the general heat flux vector (e.g. Fourier laws). Equilibrium thermodynamics implies that the variables temperature, pressure and specific volume cannot be independent, but must be related by some equation of state

$$E(w, T^{(a)}, P^{(a)}) = 0. \tag{2.4}$$

A central idea in the development of statistical mechanics has been the use of a phase variable or "order" parameter which serves to distinguish the two phases. An order parameter for a particular material or system is a variable which can be related to macroscopic observables, and, in many statistical mechanical systems, is also related to microscopic quantities. In general, one expects a canonical order parameter for a particular system, so that other possible order parameters could be expressed in terms of this natural one. However, an order parameter can be vector valued, so that it consists of an arbitrary number of independent components. A general theme which has emerged from decades of statistical mechanics research is that the behavior of the order parameter obeys some general physical laws for a broad range of systems even though their respective order parameters consist of distinct physical variables.

In the most general case, we assume an order parameter $\lambda(t, x)$ which has been normalized so that λ near 0 is one phase, e.g. solid, while λ near 1 is the other, e.g. liquid. We assume the liquid-solid terminology for convenience, even though the development is for general phase transitions. We note that from a mathematical perspective, it may be aesthetically desirable to have fixed values of λ (e.g. $\lambda = 0$ and $\lambda = 1$) for the pure phases. At the same time it is of some practical importance to have a linear λ dependence in the heat balance equation. For a general system with an arbitrary phase diagram, this linear dependence and fixed values of λ are together incompatible with basic thermodynamic identities. We thus choose the linear dependence in favor of the fixed values of λ.

In general, an important distinction between order parameters is whether they are locally conserved or nonconserved. For a nonconserved order parameter, λ, the total time derivative of λ must balance the extent to which it is out of equilibrium. That is, for a total Helmholtz free energy $\mathcal{F} = \int_\Omega F/w dx$ and a rate constant A, one has the equation

$$\frac{d\lambda}{dt} = -A\mathcal{F}_\lambda. \tag{2.5}$$

Analogously, for a system which is governed by a conserved order parameter, one has the equation

$$\frac{d\lambda}{dt} = A\Delta\mathcal{F}_\lambda \tag{2.6}$$

which replaces (2.5).

The system of equations (2.1) - (2.5) represents a very general model which is complete within the context of the problem analyzed and can be studied upon specifications of the form of $E, \vec{\vec{P}}, \vec{Q}$ and F. In order to perform further analysis which will lead to specific macroscopic results, one must make suitable choices of these functions which are nevertheless sufficiently general to encompass a broad spectrum of materials. This is done below in Section 3.

3. The derivation. We develop first the equation of state (2.4) by considering a point (P_E, T_E) on the coexistence curve of the equilibrium phase diagram (see Figures 1, 2). It follows from general thermodynamics that the pressure and temperature (P_E, T_E) uniquely determine the specific volume w_E, $[= w_E(\lambda)]$ in each phase, denoted $w_E(1), w_E(0)$, respectively, for liquid and solid. We use the notation

$$\bar{w} \equiv \frac{1}{2}\{w_E(1) + w_E(0)\}; \qquad [w]_E \equiv w_E(1) - w_E(0) \tag{3.1}$$

and similarly for other variables. Near the point (P_E, T_E), one may represent $w_E(\lambda)$ in a series expansion in terms of T and P (the deviation from T_E and P_E) by using a small parameter ε which will scale the coefficients (see Figure 3). We define

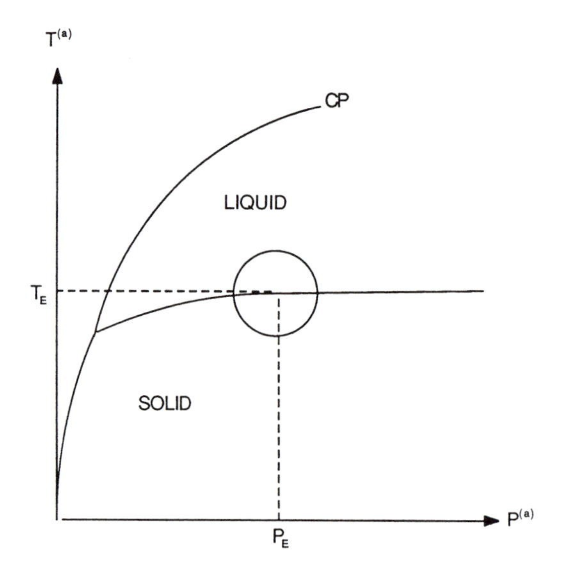

Figure 1: Coexistence curve in the pressure-temperature plane. Local expansions are performed about an arbitrary point (P_E, T_E) on the liquid-solid boundary, away from the triple point.

$$\begin{aligned} \varepsilon\eta &\equiv \frac{\partial w}{\partial T}\Big|_{P,\lambda} = \text{coefficient of thermal expansion,} \\ \varepsilon^2\nu &\equiv -\frac{\partial w}{\partial P}\Big|_{T,\lambda} = \text{coefficient of isothermal compressibility} \end{aligned} \tag{3.2}$$

Neglecting terms of order T^2 and P^2, the equation of state may be approximated by

$$w(T, P, \lambda) = w_E(\lambda) + \varepsilon\eta(\lambda)T - \varepsilon^2\nu(\lambda)P \tag{3.3}$$

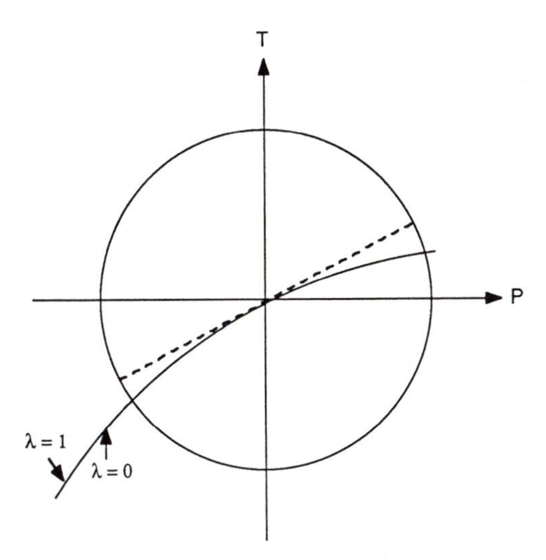

Figure 2: Magnification of the region in Figure 1 where the local approximations are made. A linear approximation is employed for the coexistence curve.

Next, the free energy is assumed to be of Landau-Ginzburg type with a term $q(T, w, \lambda)$ to incorporate the additional physics, i.e.

$$F = \frac{1}{2}\frac{D}{A}|\nabla\lambda|^2 + \frac{\alpha}{4}\lambda^2(1-\lambda)^2 + q(T, w, \lambda) \tag{3.4}$$

A natural constraint which is imposed on q, namely,

$$q(0, w_s, 0) = q(0, w_\ell, 1) = q_\lambda(0, w_s, 0) = q_\lambda(0, w_s, 1), \tag{3.5}$$

(with $w_s \equiv w_E(0)$, etc.) is needed to guarantee the correct thermodynamics of the pure phases.

There are two basic thermodynamic identities which are necessary for the compatibility of (3.3) and (3.4), namely,

$$-(P + P_E) = F_w\big|_{T,\lambda} = q_w \qquad -S = F_T\big|_{w,\lambda} = q_T \tag{3.6a, b}$$

together with an identity for the specific heat, $C(\lambda)$ (per mass), i.e.,

$$C(\lambda) = (T + T_E)S_T\big|_{w,\lambda}, \tag{3.7}$$

which are instrumental in deriving an expression for q.

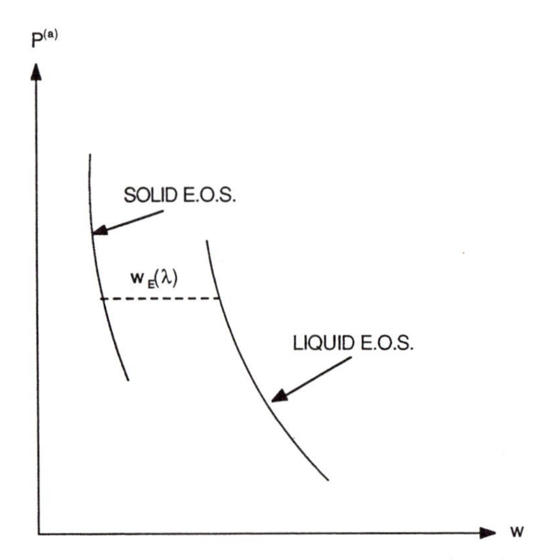

Figure 3: The coexistence plateau in the specific volume pressure plane. The equilibrium specific volume $w_E(\lambda)$ sweeps out an isobar on the plateau.

Using (3.3) in the left hand side of (3.6a) one obtains

$$q_w = \nu^{-1}\varepsilon^{-2}(w - w_E) - \varepsilon^{-1}\frac{\eta}{\nu}T - P_E. \tag{3.8}$$

Upon integration with respect to w one has

$$q = \frac{\nu^{-1}\varepsilon^{-2}}{2}w^2 - (P_E + \nu^{-1}\varepsilon^{-1}w_E + \varepsilon^{-1}\frac{\eta}{\nu}T)w + q_1(T,\lambda) \tag{3.9}$$

where $q_1(T,\lambda)$ is the integration constant which is yet to be determined.

By differentiating the free energy with respect to temperature and using (3.6b) with (3.9) the identity

$$-S = q_{1T} - \frac{\varepsilon^{-1}\eta}{\nu}w. \tag{3.10}$$

is obtained. Note that w and λ are held fixed in this differentiation. Combining (3.10) with the thermodynamic identity (3.7), one obtains

$$C(\lambda) = -(T + T_E)q_{1TT} \tag{3.11}$$

so that successive integration with respect to temperature results in the expressions

$$q_{1T} = -C(\lambda)\ell n(T + T_E) + q_2(\lambda), \tag{3.12}$$

$$q_1 = -C(\lambda)[\ell n(T + T_E) - 1](T + T_E) + q_2(\lambda)T + q_3(\lambda). \tag{3.13}$$

By substituting (3.12) into (3.10) and settling all conditions at equilibrium, i.e. $T = 0, P = 0, S = S_E, w = w_E$ one has

$$S_E(\lambda) = \frac{\varepsilon^{-1}\eta}{\nu} w_E(\lambda) + C\ell n T_E - q_2(\lambda). \tag{3.14}$$

Since this expression determines $q_2(\lambda)$, we can substitute it into the full expression for q, i.e. (3.9) with (3.13) used for q_1. We thus have an expression for q which has one arbitrary function of λ. By rearranging terms so that q has the form of an expansion about equilibrium, one obtains

$$\begin{aligned} q(T, w, \lambda) \equiv & \frac{\nu^{-1}\varepsilon^{-2}}{2}(w - w_E)^2 - P_E(w - w_E) - \frac{\varepsilon^{-1}\eta}{\nu}(w - w_E)T \\ & - C\{(T + T_E)\ell n(1 + T/T_E) - T\} - S_E(\lambda)T + q_E(\lambda), \end{aligned} \tag{3.15}$$

so that

$$q_T\Big|_{w,\lambda} = -\frac{\varepsilon^{-1}\eta}{\nu}(w - w_E) - S_E + C\ell n(1 + T/T_E), \tag{3.16}$$

where the remainder term is now relabeled $q_E(\lambda)$ since it is the value of q when T and w are at equilibrium values (i.e. $T = 0, w = w_E$).

We are now in a position to derive a more specific form of the heat balance equation by substituting (3.16) into (3.6b) which is then used in the left hand side of (2.3). One obtains the general equation

$$(T+T_E)\{\varepsilon^{-1}\frac{d}{dt}(\frac{\eta}{\nu}(w-w_E))+\frac{d}{dt}(C\ell n(1+T/T_E))+\frac{d}{dt}S_E(\lambda)\}$$

$$=w\vec{\nabla}\cdot\vec{Q} \tag{3.17}$$

By specifying the Fourier law $\vec{Q}=k(T,\lambda)\vec{\nabla}T$ and performing the differentiation in (3.17), one obtains upon using the equation of continuity (2.1) the heat balance equation

(3.18)

$$\{C'(\lambda)\ell n(1+T/T_E)+\varepsilon^{-1}(\frac{\eta}{\nu})'w-\varepsilon^{-1}(\frac{\eta}{\nu}w_E)'+S_E'\}(T+T_E)\frac{d\lambda}{dt}$$
$$+C(\lambda)\frac{dT}{d\lambda}+\varepsilon^{-1}\frac{\eta}{\nu}(T+T_E)w\vec{\nabla}\cdot\vec{u}=w\nabla\cdot k\nabla T$$

The equation above is valid for arbitrary functions $w_E(\lambda), S_E(\lambda), C(\lambda), \eta(\lambda), \nu(\lambda)$. In order to further simplify the heat balance equation, we assume a linear form for these functions, so that

$$w_E(\lambda)=\bar{w}+[w]_E(\lambda-\frac{1}{2});\qquad S_E(\lambda)=\bar{S}+[S]_E(\lambda-\frac{1}{2});$$
$$\nu(\lambda)=\bar{\nu}+[\nu]_E(\lambda-\frac{1}{2});\ etc. \tag{3.19}$$

subject to the thermodynamic identity

$$[S]_E[w]_E+P_E[\varepsilon\eta]_E[w]_E-P_E^2[\varepsilon^2\nu]_E[S]_E=0. \tag{3.20}$$

We note that once $w_E(\lambda)$ is assumed to have the form given by (3.19), $S_E(\lambda)$ must also have this form. This follows from the Clausius-Clapeyron formula for the coexistence curve

$$\frac{P}{T}\cong\frac{dP}{dT}=\frac{S_E'(1)}{w_E'(1)}=\frac{S_E'(0)}{w_E'(0)}=\frac{[S]_E}{[w]_E} \tag{3.21}$$

and effectively determines the order parameter. With the choice of (3.19) the heat balance equation becomes

$$C(\lambda)\frac{dT}{d\lambda}+\varepsilon^{-1}\frac{\eta}{\nu}(T+T_E)w\vec{\nabla}\cdot\vec{u}+$$
$$\{[C]_E\ell n(1+T/T_E)+\varepsilon^{-1}\frac{[\eta]_E\nu-\eta[\nu]_E}{\nu^2}(w-w_E)$$
$$-\varepsilon^{-1}\frac{\eta}{\nu}[w]_E+[S]_E\}(T+T_E)\frac{d\lambda}{dt}=w\vec{\nabla}\cdot k\nabla\vec{T} \tag{3.22}$$

since $C'=[C]_E$, etc. At this point we suppose that $[C]_E=0$, i.e. no jump in heat capacity, so that (3.19) implies that $C(\lambda)$ is now replaced by the constant C_E. Hence, we can write the identity (3.20) in the form

$$[w]_E=\varepsilon^2P_E[\nu]_E-\varepsilon^3\frac{P_E^3[\nu]_E[\eta]_E}{[S]_E} \tag{3.23}$$

Using (3.19), one can write $w - w_E$ as

$$w - w_E = w - \bar{w} - [w]_E(\lambda - \frac{1}{2}) \tag{3.24}$$

so that the terms involving w in (3.22) can be simplified to

$$\begin{aligned} C_E\frac{dT}{dt} &+ \varepsilon^{-1}\frac{\eta}{\nu}(T+T_E)w\vec{\nabla}\cdot\vec{u} + \\ &\{\varepsilon^{-1}(\frac{[\eta]_E}{\nu} - \frac{\eta[\nu]_E}{\nu^2})(w-\bar{w}) + [S]_E - \varepsilon P_E\frac{[\nu]_E[\eta]_E}{\nu}(\lambda - \frac{1}{2}) \\ &- \varepsilon P_E\frac{[\nu]_E}{\nu}\eta\}(T+T_E)\frac{d\lambda}{dt} \end{aligned} \tag{3.25}$$

Retaining only order one term, one has the following simplified form of the heat balance equation (2.3):

(3.26)

$$\begin{aligned} C_E\frac{dT}{dt} &+ \varepsilon^{-1}\frac{\eta}{\nu}(T+T_E)w\vec{\nabla}\cdot\vec{u} \\ &+ \{(\varepsilon\nu)^{-1}([\eta]_E - \frac{\eta}{\nu}[\nu]_E)(w-\bar{w}) + [S]_E\}(T+T_E)\frac{d\lambda}{dt} = w\vec{\nabla}\cdot k\vec{\nabla T}. \end{aligned}$$

Having specified the heat balance equation completely, we now turn attention to the phase equation (2.5). In general, a "chemical potential" conjugate to λ is defined by means of the equation

$$dF = -SdT - Pdw + \mu d\lambda \tag{3.27}$$

for $D = 0$, [cf. Eqn (3.4)], so that μ may be defined by

$$\mu \equiv \frac{dF}{d\lambda}\Big|_{T,w}. \tag{3.28}$$

When $D \neq 0$ the differentials must be replaced by variations ($d \longrightarrow \delta$). This leads to the definition

$$\tilde{\mu} \equiv \int_\Omega \mu w^{-1}dV = \frac{\delta}{\delta\lambda}\int_\Omega Fw^{-1}dV\Big|_{T,w}, \tag{3.29}$$

where $\delta/\delta\lambda$ denotes the first variation with respect to λ, for fixed w, T. The arbitrary region $\Omega(t)$ encloses a fixed mass of fluid, so that no fluid is transported through the boundary $\partial\Omega$. This means that the variation is performed subject to the constraint

$$\int_\Omega w^{-1}dV = \text{constant},$$

i.e. the mass measure $w^{-1}dV$ is stationary under the variation. The time dependence of $\Omega(t)$ is considered by use of the transport theorem,

$$\frac{d}{dt}\int_{\Omega(t)} \lambda w^{-1}dV = \int_{\Omega(t)} \frac{d\lambda}{dt}w^{-1}dV. \tag{3.30}$$

The details are presented in the Appendix. If the free energy has the form [as in (3.4)]

$$F = \frac{1}{2}\frac{D}{A}|\nabla\lambda|^2 + f(T, w, \lambda)$$

then (3.29) implies the equation

$$\mu = \frac{-w}{A}\vec{\nabla} \cdot \frac{D}{w}\vec{\nabla}\lambda + f_\lambda \tag{3.31}$$

The local equilibrium condition is $\mu = 0$ and the canonical form of the dynamical equation is given by

$$\frac{d\lambda}{dt} = -A\mu, \quad \text{or equivalently,} \quad \frac{d\lambda}{dt} = -A\frac{\delta}{\delta\lambda}\int_\Omega Fw^{-1}dV \tag{3.32}$$

While (3.31) is valid for a general f, we now specify f to be of the form indicated by (3.4) in order to establish identities which will determine $q_E(\lambda)$ and complete the specification of the free energy via (3.4) and (3.15).

We note that the equilibria of f, i.e. values $\lambda_\ell(T, P)$ and $\lambda_s(T, P)$ which solve

(3.33)

$$\begin{aligned} 0 = f_\lambda(T, w, \lambda(T, P)) &= \frac{\partial}{\partial\lambda}\{\frac{\alpha}{4\varepsilon}\lambda^2(1-\lambda)^2 + q(T, w, \lambda)\} \\ &= -\alpha\lambda(1-\lambda)(\lambda - \frac{1}{2}) + q_\lambda(T, w, \lambda)\big|_{T,w}. \end{aligned}$$

deviate from 1 and 0 by order T or P. Note that the equilibria λ_ℓ, λ_s are exactly 1 or 0 when the physical system is in equilibrium, i.e. $T = P = 0$. The overall strategy is to establish identities for $q_E(\lambda)$ analogous to (3.5) by (i) differentiating q in (3.15) with respect to λ with T and w held fixed, while w_E and S_E depend on λ, then (ii) using expansions about the equilibrium values (i.e. small T, P deviations from the point on the equilibrium phase diagram) and finally, (iii) using (3.19) to obtain the required relations.

Noting that the equation of state (3.3) implies $w - w_E = O(T)$, one sees that the imposition of conditions (3.5) on (3.15) leads to the zeroth order (in T and P, <u>not</u> in ε) conditions,

$$q_E(0) = q_E(1) = 0. \tag{3.34}$$

By differentiating q in (3.15) as indicated above and imposing the latter two conditions of (3.5) one obtains, to zeroth order, the identities

$$q_E'(1) = -P_E w_E'(1), \qquad q_E'(0) = -P_E w_E'(0). \tag{3.35}$$

Using (3.19) we see that $w_E'(\lambda) = [w]_E$ for all values of λ, so that (3.35) implies

$$q_E'(0) = q_E'(1) = -P_E[w]_E. \tag{3.36}$$

We emphasize that the functions $q_E(\lambda), w_E(\lambda), \nu(\lambda)$ etc. describe the microscopic structure of the interfacial region at a point on the coexistence curve. The simplifying assumptions being made here are based on the observation that macroscopic observables (such as interface velocity) depend only on a finite number of functionals over the interfacial structure (e.g. (4.21)). As long as critical qualitative features (such as height of the potential barrier) are maintained, the details of the internal structure are mostly irrelevant. The principle exception to this statement occurs in a neighborhood of the two equilibria (the bottoms of the wells), where macroscopic variables may indeed be sensitive to small variations in the shape and state-dependence of the potential. It is precisely these critical areas that we are able to treat quantitatively by local approximations.

We thus choose the simplest form for $q_E(\lambda)$ compatible with (3.36) and obtain

$$q_E(\lambda) \equiv a + b(\lambda - \frac{1}{2}) + c(\lambda - \frac{1}{2})^2 + d(\lambda - \frac{1}{2})^3. \tag{3.37}$$

Applying (3.34) and (3.36) one readily obtains the polynomial

$$q_E(\lambda) = P_E[w]_E(\lambda - \frac{1}{2})\{1 - 4(\lambda - \frac{1}{2})^2\}, \tag{3.38}$$

$$q'_E(\lambda) = \frac{1}{2}P_E[w]_E\{1 - 12(\lambda - \frac{1}{2})^2\} \tag{3.39}$$

The function q_λ can now be obtained by differentiating (3.15) and using the relations (3.19) along with (3.39) as

$$\begin{aligned} q_\lambda = & -\frac{1}{2}(\nu\varepsilon)^{-2}[\nu]_E(w - w_E)^2 - \nu^{-1}\varepsilon^{-2}(w - w_E)[w]_E - P_E[w_E] \\ & - \varepsilon^{-1}(\frac{\eta}{\nu})T(w - w_E) - \varepsilon^{-1}\frac{\eta}{\nu}T[w]_E - \varepsilon^{-1}(\frac{\eta}{\nu})'T(w - w_E) \\ & - [S]_E T + q'_E. \end{aligned} \tag{3.40}$$

Noting that our scaling (in terms of ε) implies $\eta = O(1), \nu = O(1), [s]_E = O(1)$ but $[w]_E = O(\varepsilon^2)$.

The last of these follows from the thermodynamic identity (3.20). We note also that the scaling of ν and η ensures that the sound speed is an increasing function of density.

From the equation of state (3.3) we see that $w - w_E = O(\varepsilon)$ while (3.20) implies

$$w - \bar{w} = w - w_E + O(\varepsilon^2) = O(\varepsilon) \tag{3.41}$$

so that $w - \bar{w}$ and $w - w_E$ can be interchanged at the $O(1)$ level. In view of these comments and the equation of state (3.3), we obtain

$$q_\lambda \cong -\frac{1}{2}\frac{[\nu]_E}{\nu^2}\eta^2 T^2 - \frac{[\eta]_E\nu - [\nu]_E\eta}{\nu^2}\eta T^2 - [S]_E T + O(\varepsilon). \tag{3.42}$$

Substituting this expression into (3.31) we obtain the new phase field equation

$$\frac{d\lambda}{dt} = A\{\frac{w}{A}\vec{\nabla}\cdot\frac{D}{w}\vec{\nabla}\lambda + \varepsilon^{-1}\alpha\lambda(1-\lambda)(\lambda-\frac{1}{2}) + \frac{\eta}{\nu}([\eta]_E - \frac{\eta}{2\nu}[\nu]_E)T^2 + [S]_E T\} \tag{3.43}$$

Coupled with equations (2.1), (3.3), (3.26) and the simplified form of Euler's equation, (2.2), namely

$$\frac{d\vec{u}}{dt} = -w\vec{\nabla}P, \tag{3.44}$$

in which the stress tensor has been replaced by pressure, one obtains a complete system of equations for $T, \lambda, \vec{u}, w$ and P. The interface for this system is now given explicitly by the set of points for which $\lambda = \frac{1}{2}$ and has thickness of order ε.

Other treatments of mechanical and thermodynamic considerations in sharp interface problems include [10 - 14].

4. A new interface relation in the sharp interface limit. We consider the phase equation (4.5) within a general geometric setting and perform a matched asymptotic expansion with the primary aim of obtaining an interface relation which generalizes (1.5), and thereby incorporates the effects of fluid properties. Upon scaling the rate constant A by writing it as $A = B_1/\varepsilon$, we have the system

$$\frac{dw}{dt} = w\vec{\nabla}\cdot\vec{u} \tag{4.1}$$

$$\frac{d\vec{u}}{dt} = -w\nabla\vec{P} \tag{4.2}$$

$$w = w_E + \varepsilon\eta T - \varepsilon^2\nu P \tag{4.3}$$

$$C_E\frac{dT}{dt} + \varepsilon^{-1}\frac{\eta}{\nu}(T+T_E)w\vec{\nabla}\cdot\vec{u} + \{(\varepsilon\nu)^{-1}([\eta]_E - \frac{\eta}{\nu}[\nu]_E)(w-\bar{w}) + [S]_E\}(T+T_E)\frac{d\lambda}{dt} = \vec{\nabla}\cdot k\nabla\vec{T} \tag{4.4}$$

$$\frac{\varepsilon^2}{\alpha B}\frac{d\lambda}{dt} = \frac{\varepsilon^2}{\alpha}\frac{D}{B}\Delta\lambda + \frac{\varepsilon^2 w}{\alpha B}\vec{\nabla}(\frac{D}{w})\cdot\vec{\nabla}\lambda + \lambda(1-\lambda)(\lambda-\frac{1}{2}) + \frac{\varepsilon}{\alpha}\{\frac{\eta}{\nu}([\eta]_E - \frac{\eta}{2\nu}[\nu]_E)T^2 + [S]_E T.\} \tag{4.5}$$

We perform an inner and outer expansion of (4.5) in the manner of [15]. The coefficient of $\Delta\lambda$ sets a natural length scale ε_1 defined by

$$\varepsilon_1^2 \equiv \frac{\varepsilon^2 D}{\alpha B}. \tag{4.6}$$

We define a new coordinate system in which the variable r is defined relative to $\Gamma(t) = \{x : \lambda(t,x) = \frac{1}{2}\}$, where r is positive when $\lambda > \frac{1}{2}$ (i.e. toward the liquid phase), with unit normal $\hat{r}_0$. The tangential variables will be of higher order and are insignificant for the derivation. Using (4.6) we take $\rho \equiv r/\varepsilon_1$ as the scaled coordinate, while $\hat{\lambda}, \hat{T}$ are the inner variables defined by

$$\hat{\lambda}(t,\rho,\cdots) = \lambda(t,r,\cdots), \qquad \hat{T}(t,\rho,\cdots) = T(t,r,\cdots), \quad \text{etc.} \tag{4.7}$$

Then (4.5) can be written as

(4.8)
$$\frac{\varepsilon_1}{D}\{-v + (\vec{u}\cdot\hat{r}_0)\}\hat{\lambda}_\rho = \hat{\lambda}_{\rho\rho} + \varepsilon_1\kappa\hat{\lambda}_\rho - \frac{1}{\hat{w}}\hat{w}_\rho\hat{\lambda}_\rho + \hat{\lambda}(1-\hat{\lambda})(\hat{\lambda} - \frac{1}{2})$$
$$+ \sqrt{\frac{B}{\alpha D}}\varepsilon_1\{\frac{\eta}{\nu}([\eta]_E - \frac{\eta}{2\nu}[\nu]_E)\hat{T}^2 + [S]_E\hat{T}\} + O(\varepsilon^2)$$

where κ is the sum of principal curvatures at the point on the interface. The mathematical subtleties involving definitions of r and κ have been treated thoroughly in [16] and are essentially the same for this problem.

We seek inner solutions of the form

$$\hat{\lambda}(t,\rho,\cdots;\varepsilon_1) = \hat{\lambda}^\circ(t,\rho,\cdots) + \varepsilon_1\hat{\lambda}^1(t,\rho,\cdots) + \cdots \tag{4.9}$$

which are to be matched with an expansion of the outer solution

$$\lambda(t,r,\cdots;\varepsilon_1) = \lambda^\circ(t,r,\cdots) + \varepsilon_1\lambda^1(t,r,\cdots) + \cdots. \tag{4.10}$$

These two expansions are to be matched in order that one obtains an approximation which is formally valid in both the regime near the interface and away from it.

The $O(1)$ inner problem consists of retaining those terms in (4.8) which have coefficients independent of ε_1, namely,

$$0 = \hat{\lambda}^\circ_{\rho\rho} + \hat{\lambda}^\circ(1-\hat{\lambda}^\circ)(\hat{\lambda}^\circ - \frac{1}{2}). \tag{4.11}$$

In addition to the trivial constant solutions $\hat{\lambda}^\circ = 0, \frac{1}{2}, 1$, equation (4.11) has the transition layer solution given by

$$\hat{\lambda}^\circ(\rho) = \frac{1}{2}\{\tanh(\rho/2) + 1\} = \frac{1}{2}\{\tanh(r/2\varepsilon_1) + 1\} \equiv \lambda(r) \tag{4.12}$$

Defining $f(\hat{\lambda}) \equiv \hat{\lambda}(1-\hat{\lambda})(\hat{\lambda} - \frac{1}{2})$, the $O(1)$ inner problem can be written as

$$\hat{\lambda}^\circ_{\rho\rho} + f(\hat{\lambda}^\circ) = 0. \tag{4.11$'$}$$

Upon defining the linear operator

$$L\hat{\lambda} \equiv \hat{\lambda}_{\rho\rho} + f'(\hat{\lambda}^\circ)\hat{\lambda}, \tag{4.13}$$

one observes from differentiating (4.11′) that $\hat{\lambda}^\circ_\rho$ is solution to the homogeneous problem $L\hat{\lambda}^\circ_\rho = 0$. We use this fact in considering the $O(1)$ problem below.

The $O(\varepsilon)$ inner problem is obtained by substituting (4.9) into (4.8) and retaining terms which are of order ε. Using (4.13) one may express this as

$$\begin{aligned} L\hat{\lambda}^1 =& -\kappa\hat{\lambda}^\circ_\rho + \frac{1}{D}(-v + (\vec{u}\cdot\hat{r}_0))\hat{\lambda}^\circ_\rho + \frac{1}{\hat{w}}\frac{\hat{w}_\rho}{\varepsilon_1}\hat{\lambda}_\rho \\ & - \sqrt{\frac{B}{\alpha D}}\{\frac{\eta}{\nu}([\eta]_E - \frac{\eta}{2\nu}[\nu]_E)(\hat{T}^\circ)^2 + [S]_E\hat{T}^\circ\} \equiv H \end{aligned} \tag{4.14}$$

Since $\hat{\lambda}^\circ_\rho$ solves the homogeneous problem, a necessary condition for the existence of a solution to (4.14) is the orthogonality of H to $\hat{\lambda}^\circ_\rho$, i.e.

$$0 = (\hat{\lambda}^\circ_\rho, H) \equiv \int_{-\infty}^{\infty} H\hat{\lambda}^\circ_\rho d\rho. \tag{4.15}$$

In simplifying this expression, we note that if T is a continuous function in the original space variables, then the matching of the $O(1)$ inner and outer solutions implies (to leading order)

$$\int_{-\infty}^{\infty} \hat{\lambda}^\circ_\rho \hat{T}(\rho)d\rho = \hat{T}(0)\int_{-\infty}^{\infty} \hat{\lambda}^\circ_\rho(\rho)d\rho = \hat{T}(0) = T(0) \tag{4.16}$$

Then (4.15) implies the new interface relation (to leading order)

$$\begin{aligned} &\{\kappa + \frac{1}{D}(v - (\vec{u}\cdot\hat{r}_0))\}\sqrt{\frac{\alpha D}{B}} \parallel \hat{\lambda}^\circ_\rho \parallel^2 \\ & - \sqrt{\frac{\alpha D}{B}}\int_{-\infty}^{\infty} \frac{1}{\hat{w}^0}\frac{\hat{w}^0_\rho}{\varepsilon_1}\hat{\lambda}^0_\rho d\rho + \frac{\eta}{\nu}([\eta]_E - \frac{\eta}{2\nu}[\nu]_E)T^2(0) + [S]_E T(0) = 0 \end{aligned} \tag{4.17}$$

where $\parallel \hat{\lambda}^\circ_\rho \parallel^2 \equiv \int_{-\infty}^{\infty}(\hat{\lambda}^\circ_\rho)^2 d\rho$.

The $O(1)$ outer solution to (4.5) is given by the roots of $f(\lambda) = 0$, namely $\lambda = 0, \frac{1}{2}, 1$. Under our assumption that r is positive into the liquid, the $O(1)$ outer solution must be

$$\lambda^\circ = \begin{cases} 0 & r < 0 \\ 1 & r > 0 \end{cases} \tag{4.18}$$

in order to satisfy the standard matching condition,

$$\lim_{r\to 0\pm} \lambda^\circ(t, r, \cdots) = \lim_{\rho\to\pm\infty} \hat{\lambda}^\circ(t, \rho, \cdots) \tag{4.19}$$

It is clear from (4.18) that the function given by (4.12) approximates the inner solution near the interface and the outer solution away from the interface.

A definition which is well-established in the study of a broad spectrum of interface problems is the interfacial free energy,

$$\sigma \equiv \frac{\mathcal{F}\{\lambda^\circ\} - \frac{1}{2}\mathcal{F}\{0\} - \frac{1}{2}\mathcal{F}\{1\}}{\text{Interface Area}}. \tag{4.20}$$

This is the free energy attributable to the interface in that it is the difference in free energy between the system with an interface minus the average of the free energies in the pure phases (with no interface). While this quantity is loosely called the "surface tension" in papers which do not consider mechanical properties, it is in fact only part of the actual surface tension when these properties are considered. The calculation of σ is identical to the original (see p. 239 of [9] or p. 5890 of [2]) with the result

$$\sigma \equiv \sqrt{\frac{\alpha D}{B}} \parallel \hat{\lambda}^\circ_\rho \parallel^2 . \tag{4.21}$$

Hence, the relation (4.17) can be written as

$$[S]_E T + \frac{\eta}{\nu}([\eta]_E - \frac{\eta}{2\nu}[\nu]_E)T^2 = -\sigma\kappa - \frac{\sigma}{D}(v - u_n) + \sigma\langle\frac{w_r}{w}\rangle_s \tag{4.22}$$

where $u_n \equiv \vec{u} \cdot \hat{r}_0$ is the component of the fluid velocity which is normal to the interface. The average $\langle f \rangle_s$ is defined with respect to the natural measure in this problem as

$$\langle f \rangle_s \equiv \frac{(f, \hat{\lambda}^0_\rho)}{(\hat{\lambda}^0_\rho, \hat{\lambda}^0_\rho)}. \tag{4.23}$$

Hence, the last term in (4.22) represents the contribution due to the gradient in specific volume across the interface. This relation is now the generalization of the Gibbs-Thomson relation and clearly reduces to (1.5) in the limit as fluid velocity and density variations approach zero. We see that σ, which plays a role in the curvature and interface velocity terms, now also influences the temperature in conjunction with the normal component of the fluid velocity, u_n and in conjunction with the change in density across the interface.

A key physical idea which emerges from this analysis is that the kinetic undercooling term is modified by replacing the front velocity with the difference between the front velocity and fluid velocity.

5. The planar equations. We consider the equations (4.1) - (4.5) in planar symmetry with the aim of deducing some simple relations between the variables. A plane wave is assumed to move with velocity v in the positive x direction under the following conditions.

(i) The functions w, λ etc. can be written as $w(x,t) = \tilde{w}(x - vt)$, etc. For convenience, we drop the tildas and let $z \equiv x - vt$.

(ii) The boundary conditions satisfy $\lambda(\pm\infty) = \lambda\pm$ where $\lambda\pm$ are the maximum and minimum roots of

$$\alpha\lambda(1-\lambda)(\lambda - \frac{1}{2}) + \varepsilon\{\frac{\eta}{\nu}([\eta]_E - \frac{\eta}{2\nu}[\nu]_E)T^2 + [S]_E T\} = 0 \tag{5.1}$$

We assume that ε is sufficiently small in order that (5.1) has three roots. Then $\lambda_+ = 1 + O(\varepsilon)$ and $\lambda_- = 0 + O(\varepsilon)$.

(iii) The front velocity is assumed to exceed the fluid velocity, i.e. $v > u$ where $\vec{u} = (u, 0, 0)$. This is a technical assumption which is valid for most situations of interest and facilitates the analysis.

We allow for general boundary conditions provided (ii) holds, and we use the symbols C_1, C_2 etc. to denote specific constants which will depend on boundary conditions. From equation (4.1) we have

$$w\frac{\partial u}{\partial x} = \frac{\partial w}{\partial t} + u\frac{\partial w}{\partial x} \quad \text{or} \quad wu' = (u-v)w' \tag{5.2}$$

which can be integrated so that

$$u = C_1 w + v. \tag{5.3}$$

The substitution of (5.3) into (4.2) then implies

$$C_1(u-v)w' = -wP' \tag{5.4}$$

Using (5.3) in (5.4) and integrating, one obtains

$$P = -C_1^2 w + C_2 \tag{5.5}$$

Having solved for u and P in terms of w, we can now solve for T as a function of w alone by using the equation of state (4.3), so that

$$T = \frac{w - w_E + \varepsilon^2\nu(-C_1^2 w + C_2)}{\varepsilon\eta} \tag{5.6}$$

The travelling wave thus imposes a simple set of relationships on the variables. In particular, the specific volume, w, is proportional to the difference between the front and fluid velocities. This is the difference which appears in the interface relation (4.22), so that the interfacial temperature is partly balanced by a specific volume term. Furthermore, relations (5.5) and (5.6) indicate that temperature and pressure are both linear functions of specific volume.

With these relations, the system of five equations reduces to two equations for λ and any one of the other variables.

Acknowledgements: It is a pleasure to thank Professor J. Glimm for very useful discussions and comments.

REFERENCES

[1] G. CAGINALP, *"Mathematical models of phase boundaries" in Material Instabilities in Continuum Problems and Related Mathematical Problems*, (Ed. J. Ball) Heriot Watt Symp. (1985-1986), pp. 35-52, Oxford Publ.

[2] G. CAGINALP, *Stefan and Hele-Shaw type models as asymptotic limits of the phase field equations*, Physical Review A 39 (1989), pp. 5887-5896.

[3] L.I. RUBINSTEIN, *The Stefan Problem*, Am. Math. Soc. Tranl. 27 American Mathematical Society, Providence R.I. (1971).

[4] J.W. GIBBS, *Collected Works*, Yale Univ. Press New Haven (1948).

[5] J.W. CAHN AND D.W. HOFFMAN, *A vector thermodynamics for anisotropic surfaces II. Curved and faceted surfaces*, Acta Metallurgica 22 (1974), pp. 1205-1214.

[6] W.W. MULLINS, *The thermodynamics of crystal phases with curved interfaces: Special case of interface isotropy and hydrostatic pressure*, Proc. Int. Conf. on Solid-Solid Phase Transformations H.I. Erinson, et.al., eds., TMS-AIME, Warrendale, PA (1983).

[7] G. HORVAY AND J.W. CAHN, Acta Met. 9 (1961), pp. 695.

[8] O.A. OLEINIK, *A method of solution of the general Stefan problem*, Sov. Math. Dokl. 1 (1960), pp. 1350-1354.

[9] G. CAGINALP, *An analysis of a phase field model of a free boundary*, Archive for Rational Mechanics and Analysis 92 (1986), pp. 205-245.

[10] J. GLIMM, *The continuous structure of discontinuities*, Proceedings of Nice Conference (Jan. 1988).

[11] V. ALEXIADES AND J.B. DRAKE, *A weak formulation for phase change problems with bulk movement due to unequal densities*, Free Boundary Problems (1990), Proc. of Montreal Conference Ed. J. Chadam.

[12] H.W. ALT AND T. PAWLOW, Free Boundary Problems (1990), Proc. of Montreal Conference Ed. J. Chadam.

[13] A. VISINTIN, *Stefan problem with surface tension*, Mathematical Models for Phase Change Problems 88 (1989), pp. 191-213.

[14] J. GREENBERG, *Hyperbolic heat transfer problems with phase transitions*, University of Maryland Preprint.

[15] G. CAGINALP, *The role of microscopic anisotropy in the macroscopic behavior of a phase boundary*, Annals of Physics 172 (1986), pp. 136-155.

[16] M.S. BERGER AND L.E. FRAENKEL, *On the asymptotic solution of a nonlinear Dirichlet problem*, J. Math. Mech. 19 (1970), pp. 553-585.

[17] I.M. GELFAND AND S.V. FOMIN, *Calculus of Variations*, Prentice-Hall, Englewood Cliffs, N.J. (1963).

Appendix

We present here some technical results on variations of multidimensional functionals with a variable domain of integration. These results are needed to properly formulate the variational problem for the equilibrium chemical potential (3.3.1), and to obtain the thermodynamic identities (3.6 a,b) for the inhomogeneous system. Consider a functional of the form

$$\mathcal{F} = \int_{\Omega} F(w, \lambda, \vec{\nabla}\lambda)\rho dV, \tag{A.1}$$

subject to the constraint

$$\mathcal{M} \equiv \int_{\Omega} \rho dV = \text{constant}. \tag{A.2}$$

Here, $\rho \equiv w^{-1}$ is the density, and λ may be interpreted as any other dependent variable. The condition (A.2) means that variations of $\mathcal{F}$ must conserve mass. Consequently, the domain of integration $\Omega \subset \mathbf{R}^3$ cannot be held fixed under the variation, but must be allowed to vary in such a way that variations of specific volume w are compensated by variations in the volume of Ω. Define

$$\begin{aligned} \rho^* &= \rho(\vec{x}) + \gamma\tilde{\rho}(\vec{x}, w) \\ \lambda^* &= \lambda(\vec{x}) + \gamma\tilde{\lambda}(\vec{x}) \\ \vec{x}^* &= \vec{x} + \gamma\vec{y}(\vec{x}, w) \end{aligned} \tag{A.3}$$

The latter equation defines a mapping of the region Ω onto the varied region Ω^*. The total variation of the mass is

$$\begin{aligned} \delta\mathcal{M} &= \frac{d}{d\gamma}\int_{\Omega^*} \rho^* dV^*|_{\gamma=0} \\ &= \int_{\Omega} (\tilde{\rho} + \rho\vec{\nabla}_T \cdot \vec{y})dV = 0, \end{aligned} \tag{A.4}$$

where $dV^* \equiv dx_1^* \wedge dx_2^* \wedge dx_3^*$ is the perturbed volume element and $\vec{\nabla}_T$ is the "total gradient", i.e.

$$\vec{\nabla}_T = \vec{\nabla} + \vec{\nabla}\rho\frac{\partial}{\partial\rho} + \vec{\nabla}\lambda\frac{\partial}{\partial\lambda} + \sum_j\vec{\nabla}\lambda_j\frac{\partial}{\partial\lambda_j}, \tag{A.5}$$

where $\lambda_j = \frac{\partial\lambda}{\partial x_j}$. Because we are varying Ω, the functions $\tilde{\rho}$ and $\vec{y}$ in (A.3) are not chosen to vanish on $\partial\Omega$ as in standard variations, but instead must be chosen to ensure that (A.4) is satisfied identically, for <u>all</u> variations of Ω. This is possible only if the variations satisfy the local conservation condition

$$\tilde{\rho} + \rho\vec{\nabla}_T \cdot \vec{y} = 0. \tag{A.6}$$

This condition is a form of mass conservation, and is the central result of this section. The variational formula (A.4) is a special case of the general formula

$$(A.7)\qquad \begin{aligned}&\delta\int_\Omega G(x,\rho,\lambda,\vec{\nabla}\lambda)dV\\ &=\int_\Omega(\frac{\partial G}{\partial\lambda}-\vec{\nabla}_T\cdot\frac{\partial G}{\partial\vec{\nabla}\lambda})(\tilde{\lambda}-\vec{y}\cdot\vec{\nabla}\lambda)+\frac{\partial G}{\partial\rho}(\tilde{\rho}-\vec{y}\cdot\vec{\nabla}\rho)\\ &+\vec{\nabla}_T\cdot(\frac{\partial G}{\partial\vec{\nabla}\lambda}(\tilde{\lambda}-\vec{y}\cdot\vec{\nabla}\lambda)+G\vec{y})dV,\end{aligned}$$

where $\frac{\partial G}{\partial\vec{\nabla}\lambda}\equiv(\frac{\partial G}{\partial\lambda_1},\frac{\partial G}{\partial\lambda_2},\frac{\partial G}{\partial\lambda_3})$. The final divergence terms in (A.7) may be converted to a boundary integral by means of the divergence theorem. The derivation of (A.7) may be obtained from [17]. We now give a direct derivation of (A.6) by noting that (A.4) combined with (A.6) is equivalent to $\delta(\rho dV)=0$ (i.e. the variation of the mass element is identically zero). Then

$$(A.8)\qquad \begin{aligned}\delta(\rho dV)&=\frac{d}{d\gamma}(\rho^* dV^*)|_{\gamma=0}\\ &=\rho\delta dV+\tilde{\rho}dV,\end{aligned}$$

where the variation of the volume element can be written as

$$(A.9)\qquad \begin{aligned}\delta dV&=\frac{d}{d\epsilon}dx_1^*\wedge dx_2^*\wedge dx_3^*|_{\gamma=0}\\ &=dy_1\wedge dx_2\wedge dx_3+dx_1\wedge dy_2\wedge dx_3\\ &+dx_1\wedge dx_2\wedge dy_3.\end{aligned}$$

But, one has $dy_1=\vec{\nabla}_T y_1\cdot(dx_1,dx_2,dx_3)$, etc. which leads to

$$(A.10)\qquad \delta(\rho dV)=(\tilde{\rho}+\rho\vec{\nabla}_T\cdot\vec{y})dV=0.$$

Finally, we apply (A.6) and (A.7) to (A.1) to get the ρ-variation

$$(A.11)\qquad \begin{aligned}\delta\mathcal{F}|_\lambda&=\int_\Omega-\frac{\partial(F\rho)}{\partial\rho}\vec{\nabla}_T\cdot(\rho\vec{y})+\vec{\nabla}_T\cdot(F\rho\vec{y})dV\\ &=\int_\Omega-\rho\frac{\partial F}{\partial\rho}(\rho\vec{\nabla}_T\cdot\vec{y}+\vec{y}\cdot\vec{\nabla}\rho)+\rho\vec{y}\cdot\vec{\nabla}_T F dV\\ &=\int_\Omega\frac{\partial F}{\partial\rho}\tilde{\rho}\rho dV,\end{aligned}$$

where $\vec{\nabla}_T=\vec{\nabla}\rho\frac{\partial}{\partial\rho}$ here. Then since

$$(A.12)\qquad \tilde{w}\equiv\delta w=-w^2\delta\rho=-w^2\tilde{\rho},$$

we get the equivalent form

$$(A.13) \qquad \delta\mathcal{F}|_\lambda = \int_\Omega \frac{\partial F}{\partial w}\tilde{w}\rho dV.$$

For a variable such as λ, no variation of the boundary is necessary, and (A.7) reduces to the standard form

$$(A.14) \qquad \begin{aligned} \delta\mathcal{F}|_w &= \int_\Omega \Big(\frac{\partial(F\rho)}{\partial\lambda} - \vec{\nabla}_T\cdot\frac{\partial(F\rho)}{\partial\vec{\nabla}\lambda}\Big)\tilde{\lambda} + \vec{\nabla}_T\cdot\Big(\frac{\partial(F\rho)}{\partial\vec{\nabla}\lambda}\tilde{\lambda}\Big)dV \\ &= \int_\Omega \Big(\frac{\partial F}{\partial\lambda} - w\vec{\nabla}_T\cdot\rho\frac{\partial F}{\partial\vec{\nabla}\lambda}\Big)\tilde{\lambda}\rho dV + \int_{\partial\Omega}\frac{\partial F}{\partial\vec{\nabla}\lambda}\cdot\vec{n}\tilde{\lambda}\rho dA, \end{aligned}$$

where now (since Ω and w are held fixed) we may assume that $\tilde{\lambda}$ vanishes on $\partial\Omega$. Formulae (A.13) and (A.14) immediately yield (3.6 a, b) and the λ variation (3.3.1).

GEOMETRIC EVOLUTION OF PHASE-BOUNDARIES

In the memory of Professor Kôsaku Yosida

YOSHIKAZU GIGA* AND SHUN'ICHI GOTO**

1. Introduction. This paper continues our study [CGG], [GG] of a motion of phase-boundaries whose speed locally depends on the normal vector field and curvature tensors.

Material science provides a lot of examples of such a motion. Let D_t denote a bounded open set in $\mathbf{R}^n$ which one phase of material, say, solid, occupies at time t. Another phase, say, liquid, occupies the region outside D_t and the two phases are bounded by an interface Γ_t called a phase boundary. We are interested in evolution of the phase boundary Γ_t. To write down the equation for Γ_t we temporary assume that Γ_t is a smooth hypersurface and Γ_t equals ∂D_t, the boundary of D_t. Let $\mathbf{n}$ denote the unit exterior normal vector field of $\Gamma_t = \partial D_t$. Let $V = V(t, x)$ denote the speed of Γ_t at $x \in \Gamma_t$ in the exterior normal direction. The equation for Γ_t we consider here is of the form

$$V = f(t, \mathbf{n}(x), \nabla \mathbf{n}(x)) \text{ on } \Gamma_t, \tag{1.1}$$

where f is a given function and $-\nabla \mathbf{n}$ is essentially the curvature tensor. For later convenience we extend $\mathbf{n}$ to a vector field (still denoted by $\mathbf{n}$) on a tubular neighborhood of Γ_t such that $\mathbf{n}$ is constant in the normal direction of Γ_t and ∇ stands for spatial derivatives in $\mathbf{R}^n$. An interesting example is

$$V = -\frac{1}{\beta(\mathbf{n})}\left(\sum_{i=1}^{n} \frac{\partial}{\partial x_i}\frac{\partial H}{\partial p_i}(\mathbf{n}) + c(t)\right) \tag{1.2}$$

which is referred to as evolution of an isothermal interface [Gu1], [Gu2], [AG]. Physically speaking $H \geq 0$ represents the interfacial energy and is defined on the unit sphere S^{n-1}; we extend H to $\mathbf{R}^n$ such that $H(\lambda p) = \lambda H(p)$, $\lambda > 0$. The function $c(t)$ represents the energy-difference between bulk phases and $\beta(\mathbf{n}) > 0$ is a kinetic constant which measures the drag opposing interfacial motion. When $H(p) = |p|, \beta(\mathbf{n}) = 1$, $c(t) = 0$ the equation (1.2) becomes

$$V = -\text{div } \mathbf{n} \tag{1.3}$$

which is often referred as the mean curvature flow equation. The equation (1.2) is interpreted as an anisotropic version of (1.3) (with driving force c).

---

*Department of Mathematics, Hokkaido University, Sapporo 060, JAPAN and IMA, University of Minnesota.

**Department of Applied Science, Faculty of Engineering 36, Kyushu University, Fukuoka 812, JAPAN.

A fundamental analytic question is to construct a solution of (1.1) for arbitrary initial data. Since solution Γ_t may develop singularities in a finite time even for (1.3) with smooth initial data [Gr], we should introduce a notion of weak solutions to track the whole evolution of Γ_t. The first attempt is done by Brakke [B] where he constructed a global varifold solution. However his solution may not be unique. Recently alternative weak formulation of solutions are introduced. The main idea is to interpret Γ_t as a level set of a function u. This idea goes back to [Se] for $n = 2$ and extended by Osher and Sethian [OS] for numerical study of

$$(1.4) \qquad V = -\text{div}\ \mathbf{n} + c \quad (c: \text{ constant }).$$

In [CGG] Y.-G. Chen and the authors introduced a weak notion Γ_t of (1.1) through viscosity solutions u in $(0, \infty) \times \mathbf{R}^n$ of the equation induced by (1.1). We constructed a *unique* global weak solution $\{\Gamma_t\}_{t \geq 0}$ with arbitrary initial data for a certain class of (1.1) including (1.2) - (1.4) as a special examples (where H is C^2 outside origin and convex, and β is continuous). In [GG] we clarify the class of (1.1) which our theory apply. Roughly speaking, our theory applies to (1.1) provided that the equation is degenerate parabolic and that f grows linearly in $\nabla \mathbf{n}$ (see [GG]). Note that this includes the case when H is convex not necessarily strictly convex for (1.2). The case when f depends on x will be discussed in [GGI] and [G] using comparison results in [GGIS]. Almost the same time as [CGG], Evans and Spruck [ES1] constructed the same unique solution but only for (1.3). Their method of construction is approximation while ours are Perron's method which applies to more general equation than (1.3). Very recently based on results in [CGG] Soner [S] recasted the definition of solutions and obtained the asymptotic behavior of weak solutions to (1.2) when $c < 0$ where c is independent of t. After we completed this work, we learned that analysis in [CGG] and [ES1] for (1.3) is extended by Ilmanen [I] on manifolds.

In this paper we discuss relation between classical solutions and our weak solutions when the former exists. In [ES1] Evans and Spruck proved that their solutions agree with the classical solution up to the time the latter exists. In this paper we extend their results to general equations (1.1) even if the equation is degenerate parabolic (see §3). In the meanwhile we construct a local classical solution to (1.1) provided that the equation is uniformly parabolic by a level surface approach in [ES2] where they discussed (1.3) only. Our generalization clarify the key ingredient of this method (see §2). In §4 we prove a result which suggests that the cone like singularity is impossible to appear.

The bibliography of [CGG], [GG] includes many references related to the mean curvature flow equation (1.3). We take this opportunity to note some other, related articles not cited there and not mentioned elsewhere in this paper. The equation (1.3) is derived formally as a singular limit of the Allen-Cahn equation. If the motion is smooth, the convergence is proved by Bronsard and Kohn [BK] and de Mottoni and Schatzman [MS] and the proof is simplified by [Ch] very recently. After we completed this work, we learned that Evans, Soner and Songanidis [ESS] proved the convergence even if Γ_t has singularities (assuming that Γ_t develops no interior). We also learned that Evans and Spruck [ES3] proved that $\partial\Gamma_t$ is countably $n-1$ rectifiable for almost all time t; however, little is known for the regularity of weak solution Γ_t even for (1.3). For the anisotropic version (1.2) less is known. S. Angenent discussed this topic when $n=2$ but H is not necessarily convex (see his article in this proceeding). We note that (1.2) is also derived as a formal limit of some Ginzburg-Landau equations [Ca]. Very recently Taylor [T] analyzed (1.2) by completely different method when H is a crystalline energy (so is not C^2). This topic is discussed also in [AG].

2. Local smooth solutions. There seems to be a couple of ways to construct a unique local-in-time smooth solution $\{\Gamma_t\}$ of

$$
\text{(2.1)} \qquad \text{(a)} \quad V = f(t, \mathbf{n}, \nabla \mathbf{n}) \text{ on } \Gamma_t,
$$
$$
\text{(b)} \quad \Gamma_{t|t=0} = \Gamma_0,
$$

where Γ_0 is a closed smooth hypersurface in $\mathbf{R}^n$ provided that the equation is uniformly parabolic. Usually one introduces coordinates to represent Γ_t to get equations of mappings parametrizing Γ_t. However, there is a lot of freedom to choose coordinates so the equations become highly degenerate. Hence we are forced to use the Nash-Moser implicit function theorem to get a local solution (cf. [Ha]). However, if we fix the parametrization of Γ_t one can apply the usual and more familiar implicit function theorem to get local solutions. Indeed, if we express Γ_t by "height" from Γ_0 we get a single uniform parabolic equation on Γ_0 for the height function. This approach is carried out by [B] (and [C]) for the mean curvature flow (plus a driving force) equation; see also [Hu] where the proof is implicit. Since the equation is considered on the manifold Γ_0, a lot of notation from differential geometry is necessary to write down the proof, although this approach is theoretically simpler. The third alternative method is introduced by Evans and Spruck [ES2] to solve the mean curvature flow equation locally. Their idea is to construct a signed distance function of Γ_t instead of Γ_t itself. Our goal in this section is to extend their method to more general equations (2.1). Our generalization will clarify the core of the argument.

For (2.1) we only assume smoothness of f and a uniform parabolicity to prove the local existence of smooth solutions which we should state below. Let S^{n-1} denote an $n-1$ dimensional unit sphere in $\mathbf{R}^n$. Let $\mathbf{M}_n$ denote the space of $n \times n$ real matrices and $\mathbf{S}_n$ denote the space of $n \times n$ real symmetric matrices equipped with the usual ordering. We often use the following projection:

$$
Q_{\overline{p}}(X) = R_{\overline{p}} X R_{\overline{p}}, \quad R_{\overline{p}} = I - \overline{p} \otimes \overline{p} \in \mathbf{S}_n, \quad \overline{p} \in S^{n-1}, \quad X \in \mathbf{M}_n,
$$

where I denotes the identity matrix (see [GG]). We now list our assumptions on f.

$$
\text{(2.2)} \qquad f : [0,T) \times S^{n-1} \times \mathbf{M}_n \to \mathbf{R} \text{ is smooth} .
$$

(2.3) There is a constant $\theta, 0 < \theta < 1$ such that

$$\lim_{\varepsilon \downarrow 0} \frac{1}{\varepsilon} \{f(t, -\overline{p}, -Q_{\overline{p}}(X + \varepsilon Y)) - f(t, -\overline{p}, -Q_{\overline{p}}(X))\}$$
$$\geq \theta \text{ trace } Q_{\overline{p}}(Y) \quad \text{if } Y \geq 0$$
$$\text{for all } (t, \overline{p}, X) \in [0, T) \times S^{n-1} \times \mathbf{S}_n, Y \in \mathbf{S}_n.$$

The second assumption can be interpreted as a uniform ellipticity of $-f$.

THEOREM 2.1. *Assume (2.2) and (2.3). Let Γ_0 be a smooth hypersurface which is a boundary of a bounded domain D_0 in $\mathbf{R}^n$. There is a positive number $T_* < T$ and a unique smooth family $\{\Gamma_t\}_{0 \leq t < T_*}$ of smooth closed hypersurfaces which solves (2.1).*

We derive an equation for the signed distance function of Γ_t satisfying (2.1a). From now on we assume that f is independent of t for simplicity since the proof can be trivially modified for f depending on t. Suppose that the desired solution Γ_t exists. Let D_t be the bounded domain surrounded by Γ_t. If a function $u(t, x) > 0$ in D_t and $u = 0$ on Γ_t, we see (2.1a) is equivalent to

$$u_t + F(\nabla u, \nabla^2 u) = 0 \text{ on } \Gamma_t \quad (u_t = \partial u / \partial t) \tag{2.4}$$

with

$$F(p, X) = -|p| f(-\overline{p}, -Q_{\overline{p}}(X)/|p|), \quad \overline{p} = p/|p| \tag{2.5}$$

provided that the gradient ∇u does not vanish on Γ_t (see [GG]). Here $\nabla^2 u$ denotes the Hessian matrix of u. If u is a signed distance function of Γ_t defined by

$$u(t, x) = \begin{cases} d(x, \Gamma_t), & x \in D_t \\ -d(x, \Gamma_t), & x \notin D_t, \end{cases}$$

then (2.4) is equivalent to

$$u_t + \widetilde{F}(\nabla u, \nabla^2 u) = 0 \quad \text{on } \Gamma_t \tag{2.6}$$

with

$$\widetilde{F}(p, X) = F(p, X) - \text{ trace } (X \overline{p} \otimes \overline{p}). \tag{2.7}$$

Indeed, since $|\nabla u| = 1$ in a tubular neighborhood of Γ_t, differentiation yields

$$\nabla^2 u (\nabla u \otimes \nabla u) = 0$$

which shows that the last term in (2.7) is not affected for the signed distance function. There is an advantage of (2.6) over (2.4) since (2.6) is no longer degenerate in the ∇u direction (see Proposition 2.6). A geometric consideration shows that

$$\nabla^2 d(x + r\mathbf{n}) = \nabla^2 d(x)(I + r\nabla^2 d(x))^{-1}, \quad x \in \Gamma_t, r: \text{ small },$$

where $d(x) = d(x, \Gamma_t)$. Therefore, for the signed distance function u solving (2.6) on Γ_t we have

$$u_t = G(u, \nabla u, \nabla^2 u)$$

in a *tubular neighborhood* of Γ_t (not only on Γ_t) with

$$G(r, p, X) = -\widetilde{F}(p, X(I - rX)^{-1}). \tag{2.8}$$

The above observation shows that Theorem 2.1 follows from the following Proposition 2.2 and Lemma 2.3.

PROPOSITION 2.2. *Suppose that f and Γ_0 satisfy the hypotheses of Theorem 2.1. Let v_0 be the signed distance function of Γ_0. For $\delta_0 > 0$ we set*

$$\Omega = \{x \in \mathbf{R}^n; |v_0(x)| < \delta_0\}, \quad Q_T = (0, T) \times \Omega.$$

Then for sufficiently small δ_0 there are $T_* > 0$ and a unique function $v \in C^\infty([0, T_*) \times \Omega) \times C^2(\overline{Q}_{T_*})$ which solves

$$v_t = G(v, \nabla v, \nabla^2 v) \text{ in } Q_{T_*} \tag{2.9}$$

$$|\nabla v| = 1 \text{ on } (0, T_*) \times \partial\Omega \tag{2.10}$$

$$v|_{t=0} = v_0 \text{ in } \Omega \tag{2.11}$$

in the classical sense, where G is defined by (2.8).

LEMMA 2.3. *If v solves (2.9) - (2.11) with $|\nabla v_0| = 1$ in Ω, then $|\nabla v| = 1$ in Q_{T_*}.*

Proposition 2.2 is based on a standard theory [LUS] of parabolic equations and an iteration argument, and it is rather straight forward extension of [ES2]. We just sketch the proof in the last part of this section. Lemma 2.3 reflects an interesting structure of (2.9) so we give its proof. The key ingredient is a generalization of a calculus identify

$$\frac{\partial}{\partial r}\left(\frac{x}{1 - rx}\right) = \frac{x^2}{(1 - rx)^2} = x^2 \frac{\partial}{\partial x}\left(\frac{x}{1 - rx}\right);$$

see the following two lemmas.

LEMMA 2.4. *For $r \in \mathbf{R}$ let $\Phi_r(X)$ be the Yosida approximation of $X \in \mathbf{M}_n$ i.e.*

$$\Phi_r(X) = X(I - rX)^{-1}.$$

Then

$$d\Phi_r(X)[Z] = (I - rX)^{-1} Z (I - rX)^{-1}, Z \in \mathbf{M}_n,$$

where $d\Phi_r(X)$ denotes the Fréchet derivative of $\Phi_r : \mathbf{M}_n \to \mathbf{M}_n$ at X.

Proof. The left hand side equals the directional derivative of Φ_r at X in the direction of Z. This implies

$$d\Phi_r(X)[Z] = \lim_{h \to 0} \left\{ (X + hZ)(I - r(X + hZ))^{-1} - (I - rX)^{-1} X \right\} h^{-1}$$
$$= \lim_{h \to 0} (I - rX)^{-1} \left\{ (I - rX)(X + hZ) - X(I - r(X + hZ)) \right\} (I - r(X + hZ))^{-1} h^{-1}$$
$$= \lim_{h \to 0} (I - rX)^{-1} Z (I - r(X + hZ))^{-1} = (I - rX)^{-1} Z (I - rX)^{-1}. \ \square$$

LEMMA 2.5. *(i) $d\Phi_r(X)[X^2] = (I - rX)^{-2} X^2 = \frac{\partial}{\partial r} \Phi_r(X)$.*
(ii) For $g : \mathbf{M}_n \to \mathbf{R}$ let $g_r(X)$ be

$$g_r(X) = g \circ \Phi_r(X) = g(\Phi_r(X)).$$

Then

$$\frac{\partial}{\partial r} g_r(X) = dg_r(X)[X^2] = \sum_{1 \le i,j,k \le n} \frac{\partial g_r(X)}{\partial X_{ij}} X_{ik} X_{ij}, \qquad X = (X_{ij}) \in \mathbf{M}_n.$$

Proof. Lemma 2.4 and

$$\frac{\partial}{\partial r} \Phi_r(X) = (I - rX)^{-1} X^2$$

yield (i). To show (ii) we use the chain rule to get

$$\begin{aligned} dg_r(X)[Z] &= dg(X_r) \circ d\Phi_r(X)[Z], \quad X_r = \Phi_r(X) \\ &= dg(X_r)[(I - rX)^{-1} Z (I - rX)^{-1}] \quad \text{(by Lemma 2.4)}. \end{aligned} \tag{2.12}$$

Applying (i) yields

$$dg_r(X)[X^2] = dg(X_r) \left[\frac{\partial}{\partial r} \Phi_r(X) \right] = \frac{\partial}{\partial r} g_r(X).$$

This proves (ii) since we have

$$dg_r(X)[X^2] = \sum_{i,j,k} \frac{\partial g_r(X)}{\partial X_{ij}} X_{ik} X_{kj}$$

by the definition of differentials. $\square$

We next show a uniformly ellipticity of $G = G(r, p, X)$ in (2.8) near $|p| = 1$ and $r = 0$. For $\delta > 0$, $K > 0$ we get

$$U = U_{\delta K} = \left\{ (r, p, X) \in \mathbf{R} \times \mathbf{R}^n \times \mathbf{S}_n ; |r| < \delta, \frac{1}{2} < |p| < 2, |X| < K \right\}, \tag{2.13}$$

where $|X|$ denotes the operator norm of the self-adjoint operator X.

PROPOSITION 2.6. *Assume (2.2) and (2.3) for f. Assume that δK is sufficiently small, say $\delta K \leq 3/8$. Then there is a constant $c, 0 < c < 1$ such that*

$$c|\xi|^2 \leq \sum_{1\leq i,j\leq n} \frac{\partial G}{\partial X_{ij}}(r,p,X)\xi_i\xi_j \leq c^{-1}|\xi|^2$$

for all $(r,p,X) \in \overline{U}_{\delta K}$, $\xi = (\xi_1, \cdots \xi_n) \in \mathbf{R}^n$.

Proof. The upper bound is trivial since $\partial G/\partial X_{ij}$ is continuous in $\overline{U}$. To get the lower bound we first observe that (2.3) implies

$$\sum_{i,j} \frac{\partial \widetilde{F}}{\partial X_{ij}} Y_{ij} \leq -\theta \text{ trace } Y \quad \text{for } Y \geq 0, \tag{2.14}$$

where $\widetilde{F}$ is defined by (2.7). Indeed from (2.3) and (2.5) it follows that

$$\begin{aligned}
\widetilde{F}(p, X+\varepsilon Y) - \widetilde{F}(p,X) &\leq -\varepsilon(\theta \text{ trace } (R_{\overline{p}} Y R_{\overline{p}}) + \text{ trace } (Y\overline{p}\otimes\overline{p})) \\
&= -\varepsilon(\theta \text{ trace } Y + (1-\theta) \text{ trace } (Y\overline{p}\otimes\overline{p})) \\
&\leq -\varepsilon\theta \text{ trace } Y \quad \text{for } Y \geq 0, \varepsilon > 0
\end{aligned}$$

since trace $(Y\overline{p}\otimes\overline{p}) \geq 0$. Here we have used $R_{\overline{p}}^2 = R_{\overline{p}}$ so that

$$\text{trace } (R_{\overline{p}} Y R_{\overline{p}}) = \text{ trace } (Y R_{\overline{p}}).$$

We thus obtain (2.14).

We apply the chain rule (2.12) to get

$$\begin{aligned}
\sum_{i,j} \frac{\partial G}{\partial X_{ij}}(r,p,X)\xi_i\xi_j &= dg_r(X)[\xi\otimes\xi], \qquad g_r(X) = -\widetilde{F}(p, X_r) \\
&= d(-\widetilde{F})(X_r) \circ d\Phi_r(X)[\xi\otimes\xi],
\end{aligned}$$

where $X_r = \Phi_r(X)$ is the Yosida approximation of X. Lemma 2.4 now yields

$$\sum_{i,j} \frac{\partial G}{\partial X_{ij}} \xi_i\xi_j = \sum_{i,j} \frac{\partial(-\widetilde{F})}{\partial X_{ij}} Y_{ij} \text{ with } Y = (I - rX)^{-1}\xi\otimes\xi(I-rX)^{-1}. \tag{2.15}$$

Since $|rX| \leq 3/8$ implies $(I-rX)^{-1} \geq 2I/5$, we have

$$Y \geq \left(\frac{2}{5}\right)^2 \xi\otimes\xi \geq 0.$$

Applying (2.14) to (2.15) yields

$$\sum_{i,j} \frac{\partial G}{\partial X_{ij}} \xi_i\xi_j \geq \frac{4\theta}{25}|\xi|^2. \quad \square$$

Proof of Lemma 2.3. Since v solves (2.9), a calculation shows that the function

$$w = |\nabla v|^2 - 1$$

solves

$$\tag{2.16} \begin{aligned} w_t = &\sum_{i,j} \frac{\partial G}{\partial X_{ij}} w_{ij} + \sum_{\ell} \frac{\partial G}{\partial p_\ell} w_\ell + 2\frac{\partial G}{\partial r} w \\ &+ 2\left[\frac{\partial G}{\partial r} - \sum_{i,j,k} \frac{\partial G}{\partial X_{ij}} v_{ik} v_{jk}\right], \end{aligned}$$

where all partial derivatives of G are evaluated at $(v, \nabla v, \nabla^2 v)$ and $w_\ell = \partial w/\partial x_\ell$, $w_{ij} = \partial^2 w/\partial x_i \partial x_j$ etc. Since $v_{kj} = v_{jk}$, from Lemma 2.5 (ii) it follows that

$$\frac{\partial G}{\partial r} = \sum_{i,j,k} \frac{\partial G}{\partial X_{ij}} v_{ik} v_{jk}.$$

The equation (2.16) now becomes

$$\tag{2.17} w_t = \sum_{i,j} \frac{\partial G}{\partial X_{ij}} w_{ij} + \sum_{\ell} \frac{\partial G}{\partial p_\ell} w_\ell + 2\frac{\partial G}{\partial r} w \text{ in } Q_{T_*}.$$

Since (2.17) is parabolic by Proposition 2.6, applying the maximum principle with (2.10) and $|\nabla v_0| = 1$ yields $w = 0$ in Q_{T_*}. This is the same as $|\nabla v| = 1$ in Q_{T_*}. □

Remark 2.7. If f in (2.1) depends on x in addition, (2.16) should be altered. Apparently, straight-forward extension of Theorem 2.1 for f depending on x is not easy. The proof of Proposition 2.2 is easily extended while there is a difficulty in the proof of Lemma 2.3. We do not get an equation like (2.17) yielding $w = 0$, when f depends on x.

We conclude this section by giving a brief sketch of the proof of Proposition 2.2; see [ES2] for the details. We set

$$h = G(v_0, \nabla v_0, \nabla^2 v_0)$$

and write the equation for $\tilde{v} = v - v_0 - th$ with the boundary conditions. The function $\tilde{v}$ can be interpreted as a small disturbance about $v_0 + th$. Since the equation linearized around $v_0 + th$ is uniformly parabolic in region we are interested in, we apply Schauder's estimates [LUS] to carry out the standard iteration procedure which yields a local classical solution $\tilde{v}$. A standard regularity theory [LUS] guarantees the higher regularity of v up to $t = 0$.

Remark 2.8. In the method just sketched there is a curious point on minimal regularity assumptions on the initial data v_0 in (2.11) to construct the local classical solution. We are forced to assume $v_0 \in C^{4+\alpha} (0 < \alpha < 1)$ guarantee that $v_0 + th$ is $C^{2+\alpha, 1+\alpha/2}$ which is the least regularity assumption to get a solution $\tilde{v} \in C^{2+\alpha, 1+\alpha/2}$ (see [LUS] for the definition). We wonder whether or not $v_0 \in C^{2+\alpha}$ is enough to get a $C^{2+\alpha, 1+\alpha/2}$ solution v of (2.9) - (2.11).

3. Consistency with weak solutions. The smooth solution constructed in §2 may collapse in a finite time. In [CGG] weak solutions for (2.1) is introduced so that one can track whole evolution of Γ_t; see also [GG]. In this section we show that the smooth solution agrees with the weak solution in the time interval where the former exists. This result is proved by [ES1] for the mean curvature flow equation (1.3). Although the basic idea is similar, our proof given below is simpler and more general than theirs because we do not approximate solutions.

We first recall the definition of weak solutions in [CGG, GG]. Let $\{(\Gamma_t, D_t)\}_{t\geq 0}$ be a family of compact sets and bounded open sets in $\mathbf{R}^n$. Suppose that for some $\alpha < 0$ there is a viscosity solution $u \in C_\alpha([0,T] \times \mathbf{R}^n)$ for

$$u_t + F(t, \nabla u, \nabla^2 u) = 0 \text{ in } (0,\infty) \times \mathbf{R}^n \tag{3.1}$$

such that zero level sets of $u(t,\cdot)$ at $t \geq 0$ equals Γ_t and that the set where $u(t,\cdot) > 0$ equals D_t. Here F is determined from f through (2.5) where t-dependence is suppressed. If $(\Gamma_t, D_t)|_{t=0} = (\Gamma_0, D_0)$ we say $\{(\Gamma_t, D_t)\}_{t\geq 0}$ is a *weak solution* of (2.1) (with initial data (Γ_0, D_0)). Here $T > 0$ is arbitrary and $v \in C_\alpha(A)$ means $v - \alpha$ is continuous with compact support in A.

It is now well-known that there is a unique (global) weak solution of (2.1) if D_0 is bounded open and $\Gamma_0 (\subset \mathbf{R}^n \setminus D_0)$ contains ∂D_0 provided that (2.1) is degenerate parabolic and f grows linearly in $\nabla \mathbf{n}$ ([CGG, Theorem 7.3], [GG, Theorem 3.8]). For example if $f = f(t,p,Z)$ satisfies

$$f : [0,\infty) \times S^{n-1} \times \mathbf{M}_n \to \mathbf{R} \text{ is continuous }, \tag{2.2$'$}$$

$$f(t, -\overline{p}, -Q_{\overline{p}}(X)) \geq f(t, -\overline{p}, -Q_{\overline{p}}(Y)) \text{ for } X \geq Y, \quad \overline{p} \in S^{n-1} \text{ and } t \geq 0 \tag{2.3$'$}$$

(which are weaker than (2.2) and (2.3)) and

$$C_{RT} = \sup\left\{\frac{\partial f}{\partial Z_{ij}}; |Z| \leq R, \overline{p} \in S^{n-1}, 0 < t < T, 1 \leq i,j \leq n\right\} < \infty, \tag{3.2}$$

then there is a unique global weak solution with given initial data. The condition (2.3$'$) means a degenerate ellipticity of $-f$. Our theory applies to (1.2) provided that H is convex and C^2 outside the origin and that $\beta : S^{n-1} \to \mathbf{R}$ and $c : [0,\infty) \to \mathbf{R}$ are continuous.

THEOREM 3.1. *Assume (2.2$'$), (2.3$'$) and (3.2) for f in (2.1). Let Γ_0 be a smooth hypersurface which is a boundary of a bounded domain D_0 in $\mathbf{R}^n$. Let $\{\Gamma_t^s\}_{0\leq t<T_0}$ be the local smooth solution of (2.1) with initial data Γ_0. Let $\{(\Gamma_t, D_t)\}_{t\geq 0}$ be the weak solution of (2.1) with initial data (Γ_0, D_0). Then $\Gamma_t = \Gamma_t^s$ for $0 \leq t < T_0$.*

If we assume (2.2), (2.3), we proved in §2 that there is a unique smooth local solution. However under (2.2$'$), (2.3$'$) we do not know the existence of unique classical solution of (2.1).

We again assume that f is independent of t for simplicity. Let v denote the signed distance function of Γ_t^s, i.e.

$$v(t,x) = \begin{cases} d(x, \Gamma_t^s), & x \in D_t^s \\ -d(x, \Gamma_t^s), & x \notin D_t^s, \end{cases}$$

where D_t^s is the bounded domain bounded by Γ_t^s. For the proof we construct a viscosity solution u of (3.1) in $(0, T_0) \times \mathbf{R}^n$ such that the signature of u agrees with v. Since Γ_t^s is a smooth solution of (2.1) we see as in §2 for each $T_1, 0 < T_1 < T_0$ there is $\delta > 0$ such that v solves

$$v_t = G(v, \nabla v, \nabla^2 v) \text{ in } R_{T_1}^\delta = \bigcup_{0<t<T_1} \{t\} \times \Omega_\delta(t) \tag{3.3}$$

with $\Omega_\delta(t) = \{x \in \mathbf{R}^n; |v(t,x)| < \delta\}$, where G is defined by (2.8).

LEMMA 3.2. *For sufficiently small δ the function v satisfies*

$$v_t + F(\nabla v, \nabla^2 v) \leq 0 \text{ in } R_{T_1}^\delta \cap \{v \leq 0\} \tag{3.4}$$

in the classical sense, where F is as in (2.5).

Proof. We take δ small so that $|v\nabla^2 v| \leq 3/8$ in $R_{T_1}^\delta$. Since $|rX| \leq 3/8$ implies

$$X(I - rX)^{-1} - X = rX^2(I - rX)^{-1} \leq 0 \text{ for } r \leq 0,$$

we see, by (3.3),

$$v_t + \widetilde{F}(\nabla v, \nabla^2 v) \leq v_t - G(v, \nabla v, \nabla^2 v) = 0 \text{ in } R_{T_1}^\delta \cap \{v \leq 0\}, \tag{3.5}$$

where $\widetilde{F}$ is defined by (2.7). Here we use the fact that (2.3′) implies the degenerate ellipticity

$$\widetilde{F}(p, X) \leq \widetilde{F}(p, Y) \quad \text{if } X \geq Y.$$

Since $|\nabla v| = 1$ implies

$$\nabla^2 v(\nabla v \otimes \nabla v) = 0,$$

(3.4) now follows from (3.5). □

LEMMA 3.3. *For sufficiently small δ there is σ_0 such that*

$$w(t,x) = e^{-\sigma t} v(t,x), \quad \sigma > \sigma_0$$

satisfies

$$w_t + F(\nabla w, \nabla^2 w) \leq 0 \text{ in } R_{T_1}^\delta \cap \{v \geq 0\}. \tag{3.6}$$

Proof. Since v solves (3.3), using

$$w_t = -\sigma e^{-\sigma t} v + e^{-\sigma t} v_t,$$
$$\widetilde{F}(\lambda p, \lambda X) = \lambda \widetilde{F}(p, X),$$

we see

$$w_t + \sigma w + \widetilde{F}(\nabla w, \nabla^2 w(I - v\nabla^2 v)^{-1}) = 0$$

or

$$(3.7)\quad w_t + F(\nabla w, \nabla^2 w) = F(\nabla w, \nabla^2 w) - F(\nabla w, \nabla^2 w(I - v\nabla^2 v)^{-1}) - \sigma w \text{ in } R^\delta_{T_1}.$$

By taking δ small, we may always assume that

$$|v\nabla^2 v| \le 3/8, \quad |\nabla^2 w| \le |\nabla^2 v| \le M \text{ in } R^\delta_{T_1}$$

with some M independent of t and x. A calculation now shows

$$|F(p,X) - F(p,Z)| \le K|X - Z|, \quad K = C_{3M,T_1}$$

for $X = \nabla^2 w, Y = \nabla^2 v, Z = X(I - rY)^{-1}$, $r = v$, where C_{RT} is as in (3.2). From (3.7) it follows that

$$w_t + F(\nabla w, \nabla^2 w) \le K|X - Z| - \sigma w.$$

Since

$$\begin{aligned} X(I - rY)^{-1} - X &= rXY(I - rY)^{-1} \\ &= v(\nabla^2 w)(\nabla^2 v)(I - v\nabla^2 v)^{-1} \end{aligned}$$

for $\sigma > \sigma_0$ with

$$\sigma_0 = K \sup_{R^\delta_{T_1}} |w(\nabla^2 v)^2 (I - v\nabla^2 v)^{-1}|$$

we have

$$\begin{aligned} w_t + F(\nabla w, \nabla^2 w) &\le (\sigma_0 - \sigma) w \\ &\le 0 \qquad\qquad \text{for } w \ge 0 \end{aligned}$$

This is the same as (3.6). □

PROPOSITION 3.4. *For each T_1, $0 < T_1 < T_0$ there are (viscosity) sub- and supersolutions w_- and w_+ of*

$$(3.8)\qquad u_t + F(\nabla u, \nabla^2 u) = 0 \textit{ in } (0, T_1) \times \mathbf{R}^n$$

such that $w_- \le w_+$ and $w_\pm$ has the same signature as v and satisfies

$$\begin{aligned} w_\pm(0,x) &= v(0,x) && \textit{if } |v| < \delta_0 \\ w_\pm(t,x) &= -\delta_0 && \textit{if } v < -\delta_0 \\ w_\pm(t,x) &= \delta_0 && \textit{if } v > \delta_0 \end{aligned}$$

for some $\delta_0 > 0$.

Proof. Let δ be as in Lemmas 3.2 and 3.3. We set

$$w_- = \begin{cases} \min(w, \delta_0) & \text{if } v \ge 0 \\ \max(v, -\delta_0) & \text{if } v \le 0 \end{cases}$$

with $\delta_0 = \delta/2$ where $t \in (0, T_1)$. Clearly $w_-(0,x) = v(0,x)$ for $|v| < \delta_0$ and w_- has the same signature as v. Since w is a subsolution of (3.8) in $\{v \geq 0\}$ and v is a subsolution in $\{v \leq 0\}$ and since F is geometric

$$\min(w, \delta_0), \quad \max(v, -\delta_0)$$

are subsolutions of (3.8) in $\{v \geq 0\}$, $\{v \leq 0\}$, respectively (see [CGG, Theorem 5.2]).

We shall prove that w_- is a subsolution of (3.8) in $(0, T_1) \times \mathbf{R}^n$. To show this it remains to show that w_- is a subsolution on the set of $v = 0$. Since Γ_t^s is smooth, it is not difficult to prove that

$$\mathcal{P}^{2,+} w_-(t,x) \subset \mathcal{P}^{2,+} w \cup \mathcal{P}^{2,+} v(t,x),$$

where $\mathcal{P}^{2,+}$ denotes the parabolic super 2-jets (see e.g. [GGIS] for the definition). On Γ_t^s we have

$$w_t + F(\nabla w, \nabla^2 w) \leq 0$$
$$v_t + F(\nabla v, \nabla^2 v) \leq 0$$

in the classical sense so for $(\tau, p, X) \in \mathcal{P}^{2,+} w_-(t,x)$ such that $v(t,x) = 0$ we see

$$\tau + F(p, X) \leq 0$$

which implies w_- is a subsolution on the set $\{v = 0\}$.

In the same way we see v and w are supersolutions in $\{v \geq 0\}$ and $\{v \leq 0\}$, respectively. We thus conclude

$$w_+ = \begin{cases} \min(v, \delta_0) & \text{if } v \geq 0 \\ \max(w, -\delta_0) & \text{if } v \leq 0 \quad (w = e^{-\sigma t} v) \end{cases}$$

is the desired supersolution; $w_- \leq w_+$ follows from the definition. □

Proof of Theorem 3.1. Using $w_\pm$ in Proposition 3.4, Perron's method [CGG, Theorem 4.5] provides a viscosity solution u of (3.8) in $(0, T_1) \times \mathbf{R}^n$ such that $w_- \leq u \leq w_+$ on $[0, T_1) \times \mathbf{R}^n$. Since u is forced to have the same signature as v, we find $\Gamma_t = \Gamma_t^s$, $D_t = D_t^s$ for $0 \leq t < T_1$. Since T_1 is an arbitrary positive number less than T_0, the proof is now complete. □

Remark 3.5. After this work was completed, Evans, Soner and Songanidis [ESS] proved that the signed distance function d of the weak solution Γ_t satisfies

$$d_t - \Delta d \leq 0 \text{ in } \{d < 0\}$$

globally in space in the viscosity sense provided that Γ_t solves the mean curvature flow equation (1.3). This can be interpreted as an improvement of Lemma 3.2 for (1.3). It would be interesting to study whether

$$d_t + \widetilde{F}(\nabla d, \nabla^2 d) \leq 0 \text{ in } \{d < 0\}$$

holds for (2.1) with general f.

4. Singularities. Smooth local solutions may collapse in a finite time even for the mean curvature flow equation (1.3). It is shown in [Gr] that a barbell with a long, thin handle actually becomes singular (see also [A]). A natural question is what kind of singularities actually appears. In [Hu] Huisken considered a two-dimensional rotational symmetric hypersurface with positive mean curvature and proved that the singularity behaves like cylinders asymptotically. Note that this does not exclude cusp singularities. The place of singularities related to this problem is discussed in [K].

Our result in this section suggests that an asymptotically cone like singularity is impossible to occur in the motion by mean curvature. We show in particular that if the singularity is nondegenerate in some sense, then it looks like sphere asymptotically.

THEOREM 4.1. *Let $\{(\Gamma_t, D_t)\}_{t>0}$ be a weak solution of (1.3). Let u be a defining function of (Γ_t, D_t) i.e.*

$$\Gamma_t = \{x \in \mathbf{R}^n; u(t,x) = 0\},$$
$$D_t = \{x \in \mathbf{R}^n; u(t,x) > 0\}$$

and u solves

$$u_t - |\nabla u| \operatorname{div}\left(\frac{\nabla u}{|\nabla u|}\right) = 0 \tag{4.1}$$

in the viscosity sense. Let $t_0 > 0$ and $x_0 \in \Gamma_{t_0}$. Suppose that

$$u(t,x) = c\tau + \frac{1}{2}\nabla^2 u(t_0,x_0)y \cdot y + 0(|y|^2 + |\tau|) \text{ as } |y|^2 + |\tau| \to 0, \tag{4.2}$$

where $\tau = t - t_0, y = x - x_0, t \le t_0$. Then u is of the form

$$u(t,x) = c\tau + \lambda \sum_{i=1}^{k} y_i^2 + o(|y|^2 + |\tau|) \tag{4.3}$$

with $c = 2(k-1)\lambda$, $\lambda \neq 0$ for some $k \le n$.

Proof. We rescale u around (t_0, x_0):

$$u_\sigma(\tau, y) = \frac{1}{\sigma^2} u(\sigma^2\tau + t_0, \sigma y + x_0), \quad \sigma > 0.$$

By (4.2) u_σ converges to a polynomial v of the form

$$v(\tau, y) = c\tau + Ay \cdot y, \quad A \in \mathbf{S}_n$$

uniformly in every compact set in $(-\infty, 0] \times \mathbf{R}^n$. By the stability [CGG, Proposition 2.4] we observe that v is a viscosity solution of (4.1) in $(-\infty, 0) \times \mathbf{R}^n$. We may assume A is diagonal by a rotation. Plugging

$$v(\tau, y) = c\tau + \sum_{j=1}^{k} \lambda_j y_j^2 \quad (\lambda_j \neq 0 \text{ for all } 1 \le j \le k)$$

in (4.1) yields $\lambda_1 = \lambda_2 = \cdots = \lambda_k$ and

$$v(\tau, y) = c\tau + \lambda \sum_{j=1}^{k} y_j^2, c = 2(k-1)\lambda$$

for some k. This is the same as (4.3). □

Remark 4.2. If $\det \nabla^2 u(t_0, x_0) \neq 0$, then $k = n$. This means singularities look like sphere, which is very related to results in [Hu]. The assumption (4.2) is fulfilled if we assume u is C^2 and $\nabla u(t_0, x_0) = 0$. However, we still do not know whether or not one can take such a defining function u near singularities. After we completed this work, we learned that Ilmanen [I] showed that u is not necessarily C^2 even if u is initially C^2.

REFERENCES

[A] S. Angenent, *Shrinking doughnuts*, Proc. of the conference on elliptic and parabolic equations held at Gregymog, Wales, August 1989.

[AG] S. Angenent and M. Gurtin, *Multiphase thermomechanics with interfacial structure. 2. Evolution of an isothermal interface*, Arch. Rational Mech. Anal. 108 (1989), pp. 323–391.

[B] K.A. Brakke, *The motion of a surface by its mean curvature*, Princeton University Press, 1978.

[BK] L. Bronsard and R. Kohn, *Motion by mean curvature as the singular limit of Ginzburg-Landau dynamics*, preprint.

[Ca] G. Caginalp, *The role of microscopic anisotropy in the macroscopic behavior of a phase boundary*, Annals of Physics, 172 (1986), pp. 136–155.

[Ch] X. Chen, *Generation and propagation of the interface for reaction-diffusion equation*, preprint.

[C] X.-Y. Chen, *Dynamics of interfaces in reaction diffusion systems*, Hiroshima Math. J. 21, to appear (1991).

[CGG] Y.-G. Chen, Y. Giga and S. Goto, *Uniqueness and existence of viscosity solutions of generalized mean curvature flow equations*, J. Differential Geometry 33, to appear (1991); (Announcement: Proc. Japan Acad. Ser. A, 65 (1989), pp. 207–210).

[ESS] L.C. Evans, H.M. Soner and P.E. Songanidis, *The Allen-Cahn equation and generalized motion by mean curvature*, manuscript.

[ES1] L.C. Evans and J. Spruck, *Motion of level sets by mean curvature I*, J. Differential Geometry, to appear.

[ES2] L.C. Evans and J. Spruck, *Motion of level sets by mean curvature II*, J. Differential Geometry, to appear..

[ES3] L.C. Evans and J. Spruck, *Motion of level sets by mean curvature III*, preprint 1990.

[GG] Y. Giga and S. Goto, *Motion of hypersurfaces and geometric equations*, J. Math. Soc. Japan, to appear.

[GGIS] Y. Giga, S. Goto, H. Ishii and M.-H. Sato, *Comparison principle and convexity preserving properties for singular degenerate parabolic equations on unbounded domains*, Indiana Univ. Math. J., to appear.

[GGI] Y. Giga, S. Goto and H. Ishii, *Global existence of weak solutions for interface equations coupled with diffusion equations*, preprint 1990.

[G] S. Goto, *A level surface approach to interface dynamics*, in preparation.

[Gr] M. Grayson, *A short note on the evolution of a surface by its mean curvature*, Duke Math. J. 58 (1989), pp. 555–558.

[Gu1] M. Gurtin, *Towards a nonequilibrium thermodynamics of two-phase materials*, Arch. Rational Mech. Anal. 100 (1988), pp. 275–312.

[Gu2] M. Gurtin, *Multiphase thermomechanics with interfacial structure. 1, Heat conduction and the capillary balance law*, Arch. Rational Mech. Anal. 104 (1988), pp. 195–221.

[Ha] R.S. Hamilton, *Three manifolds with positive Ricci curvature*, J. Differential Geometry 17 (1982), pp. 255–306.

[Hu] G. Huisken, *Asymptotic behaviour for singularities of the mean curvature flow*, J. Differential Geometry 31 (1990), pp. 285–299.

[I] T. Ilmanen, *Generalized flow of sets by mean curvature on a manifold*, preliminary report.

[K] B. Kawohl, *Remarks on quenching, blow up and dead core*, Proc. of the conference on elliptic and parabolic equations held at Gregymog, Wales, August 1989.

[LUS] O.A. Ladyzhenskaya, V. Solonnikov and N. Ural'cera, *Linear and Quasilinear Equations of Parabolic Type*, Translations of Mathematical Monographs, vol. 23, AMS, 1968.

[MS] P. DeMottoni and M. Schatzman, *Geometrical evolution of developed interfaces*, preprint; (Announcement: Evolution géometric d'interfaces, C.R. Acad. Sci. Paris, 309 (1989), pp. 453–458).

[OS] S. Osher and J.A. Sethian, *Front propagating with curvature dependent speed: Algorithms based on Hamilton-Jacobi formulations*, J. Comput. Phys. 79 (1988), pp. 12–49.

[Se] J.A. Sethian, *Curvature and evolution of fronts*, Comm. Math. Physics 101 (1985), pp. 487–499.

[S] H.M. Soner, *Motion of a set by the curvature of its boundary*, preprint 1990.

[T] J. Taylor, *Motion of curves by crystalline curvature*, preliminary report.

THE APPROACH TO EQUILIBRIUM: SCALING, UNIVERSALITY AND THE RENORMALISATION GROUP

NIGEL GOLDENFELD*

Abstract. Evidence is accumulating that the long-time behaviour of certain non-equilibrium systems shows scaling behaviour. This assertion is demonstrated in the cases of spinodal decomposition, block copolymer phase separation, crystal growth, and a particular diffusion process which can occur in porous media. These, and hopefully other non-equilibrium problems may be studied by computationally efficient numerical methods, which are based not upon discretising partial differential equations, but upon a coarse-grained description of the dynamics.

In the case of non-linear diffusion in a porous medium, it is shown that the renormalisation group can be used to study the long-time behaviour, and to calculate perturbatively the exponents characterising the anomalous diffusion.

Key words. Phase separation, crystal growth, dynamical scaling, similarity solution, renormalisation group.

1. Introduction. This contribution is concerned primarily with the *intermediate asymptotics* of the approach of a thermodynamic system to its equilibrium state[1]. This means the following. An infinite system may never reach equilibrium in a finite time, but may attain a regime in which a particular similarity solution describes the dynamics. A real system is, of course, finite; nevertheless there may well be a regime in time during which the behaviour is well-approximated by the aforementioned similarity solution for the infinite system. This is known as the intermediate asymptotic regime. Eventually, the real system will respond to the influence of the boundaries and the similarity solution will no longer describe the behaviour. The duration of the intermediate asymptotic regime depends upon the size of the system, and in practice, this regime is often easily observable. Typical examples are provided by phase separation in alloys[2], the growth of domains during the microphase segregation of a block copolymer melt[3,4], the dynamics of the superfluid transition[5], the gravitational accretion of galaxies[6] and the spread of groundwater in a porous medium[7].

This report summarises the results of a number of projects, conducted in collaboration with Fong Liu, Olivier Martin and Yoshi Oono. Since the work has been recently published and is easily available, it does not seem worthwhile to describe it at any length. Instead, I will briefly mention the principal results and the approach taken, and refer the interested reader to the original literature.

---

*Department of Physics, University of Illinois at Urbana-Champaign, 1110 West Green St., Urbana, Il. 61801. This work was partially supported by the National Science Foundation under Grant No. NSF-DMR-87-01393. Computational aspects of this work were supported by the National Science Foundation under Grant No. NSF-DMR-89-20538 administered through the Illinois Materials Research Laboratory. The author gratefully acknowledges receipt of an Alfred P. Sloan Foundation Fellowship.

2. Spinodal decomposition. Consider the quench of a binary alloy from a temperature above the coexistence curve to a point in the unstable two-phase region of the phase diagram[2]. For simplicity, we restrict ourselves to the case of a symmetric phase diagram and the case of a quench through the critical point. The equilibrium concentration of one of the components, $c(\mathbf{r}, t)$, is the quantity of interest, as it varies in time t. Experimentally, it is possible to measure the dynamic structure factor

$$S(\mathbf{k}, t) \equiv \int e^{-i\mathbf{k}\cdot\mathbf{r}} \langle c(\mathbf{r}, t)c(0,0)\rangle \, d\mathbf{r} \tag{1}$$

where the angle brackets denote an ensemble average. It is found that, at sufficiently large times, S fits the dynamical scaling form

$$S(k, t) = \ell(t)^d \Phi(k\ell(t)) \tag{2}$$

where the characteristic length is given by

$$\ell \sim t^{\phi}. \tag{3}$$

Here d is the dimension of space, the exponent ϕ is believed to be close to 1/3, and Φ is a *scaling function*, believed to be independent of the actual physical system studied. This result is supported by both computer simulation and experiment. The goal of much work in this area is to account for the existence of a *statistically self-similar* dynamics, exemplified by eqn. (2), and to calculate the scaling function Φ and the exponent ϕ.

The conventional theoretical description of spinodal decomposition in a binary alloy (BA) is the Cahn-Hilliard equation. This description has two parts: (i) a coarse-grained continuum theory for the free energy of the system, given the spatial dependence of the concentration of one of the components of the alloy, and (ii) a dynamical equation expressing the conservation of matter and Fick's Law.

Part (i): the equilibrium state of the system is found, for given boundary conditions, by minimising (subject to constant total mass of the component) the *Cahn-Hilliard free energy functional*

$$F\{c(\mathbf{r})\} = \int \left[\frac{1}{2}(\nabla c)^2 + f(c)\right] dV \tag{4}$$

where the homogeneous free energy density is given by

$$f(c) = \frac{1}{2}\tau c^2 + \frac{1}{4}uc^4. \tag{5}$$

Here, $\tau \propto T - T_c$, where T is the temperature and T_c is the critical temperature, and the coefficient u is positive. The gradient term in (1) represents the free energy associated with maintaining a concentration gradient in the system; the units have been chosen so as to make the coefficient of the gradient term equal to 1/2. The function $c(\mathbf{r})$ is assumed to have no appreciable variation on length scales much smaller than the bulk correlation length: thus F is sometimes referred to as the coarse-grained free energy. The Lagrange multiplier introduced during the course of the constrained minimisation is known as the chemical potential. The Cahn-Hilliard expression for $F\{c(\mathbf{r})\}$ reproduces the qualitative features of the phase diagram of the binary alloy. Near the critical point, higher order terms in c introduce no new corrections to the critical exponents and other universal quantities characteristic of the critical point. Furthermore, the critical behaviour is also independent of the value of u, as long as $u > 0$. These important results of *renormalisation group theory* are referred to as *universality*: many different physical systems, but differing only in the coefficients of the powers of c in $F\{c(\mathbf{r})\}$, nevertheless share the same static critical behaviour[8].

Part (ii): the dynamics of the approach of this system to its equilibrium is conventionally described by combining the equation of continuity with Fick's Law — that the mass current is proportional to $\delta F/\delta c$ — to yield the celebrated Cahn-Hilliard equation:

$$\left.\frac{\partial c}{\partial t}\right)_{BA} = M\Delta\left(\frac{\partial f}{\partial c} - \Delta c\right). \tag{6}$$

Here M is a kinetic coefficient and the subscript BA denotes binary alloy, for future reference. As mentioned above, the validity of the Cahn-Hilliard equation might be expected to be restricted to the critical region, at best. Nevertheless, it is generally used even when the quench is to a temperature well below the critical region. Later, we shall discuss why this may be reasonable.

3. Block copolymer melts. A block copolymer is a polymer chain of degree of polymerisation N, in which the first $N/2$ monomers are of molecular species A and the remaining monomers are of species B. Consider a melt of such chains. The species A and B are chosen so that below a critical temperature T_c, the spatially uniform state is unstable[3,4]. Nevertheless, phase separation is unable to proceed to completion because the A and B monomers are constrained to lie on the same polymer chain. The result is a *microphase separation*, in which domains are formed whose size is governed by the dimensions of the polymer chains, and whose shape may be lamellar, cylindrical or spherical. The initial stages of domain growth resemble spinodal decomposition, but after a sufficiently long time, the domains cease to grow, having almost reached their equilibrium dimensions. Thereafter, very slow relaxation processes anneal out defects in the domain structure, and drive the system to its complete equilibrium; however, we shall not be concerned with this very slow process here. Close to the final equilibrium of lamellar domains, the characteristic size is found to be

$$\lambda \sim N^{\theta} \tag{7}$$

where a new exponent θ has been defined. Experimentally, θ is found to be about 2/3.

4. A scaling law. The two exponents θ and ϕ describe quite different physics: the former is a property of the block copolymer system in equilibrium, the latter describes the intermediate asymptotics of spinodal decomposition. Nevertheless, it can be shown[3,4] that they are related by

$$\theta = 2\phi. \tag{8}$$

The basis of the scaling law is that the block copolymer (BCP) melt differs from the binary alloy (BA) only in that the connectivity of the chains must be taken into account. It can be shown that this leads to

$$\left.\frac{\partial c}{\partial t}\right)_{BCP} = \left.\frac{\partial c}{\partial t}\right)_{BA} - \epsilon c \tag{9}$$

where $\epsilon = \gamma/N^2$, γ is a parameter reflecting the connectivity of the chain and BCP denotes block copolymer. The *scaling law* is obtained from the *scaling hypothesis* that in both the block copolymer case and the binary alloy case, the characteristic length scale of the pattern is

$$\ell(t,\epsilon) = \epsilon^{-\theta/2} G(\epsilon t) \tag{10}$$

where G is a scaling function. For $\epsilon \neq 0$, $\ell \to \lambda$ as $t \to \infty$. Thus, the scaling function must satisfy $G(x) \to$ constant as $x \to \infty$. On the other hand, for $\epsilon = 0$, $\ell \to t^{\phi}$ as $t \to \infty$. Therefore, $G(x) \to x^{\theta/2}$ as $x \to 0$, so that the ϵ dependence of ℓ is cancelled out in the binary alloy limit. Thus, we conclude that $\theta = 2\phi$.

Is the scaling hypothesis correct? Although there is no theoretical derivation, it can be tested by numerical simulation. The dynamics of the microphase separation in block copolymer melts in two dimensions has been simulated and the characteristic size of the lamellae measured. When $\epsilon^{\theta/2}\ell(t,\epsilon)$ is plotted against ϵt for different ϵ and t, the data are found to collapse onto a single curve for $\theta = 0.65 \pm 0.02$. In this way, the scaling hypothesis is verified empirically, and the exponents θ and thence ϕ are determined.

5. Crystal growth. When a crystal of a pure substance grows into its supercooled melt, the motion of the interface is limited by the diffusion of latent heat[9]. Conventional numerical techniques are not yet developed to the point where the evolution of complex spatial patterns can be explored. Instead, we have begun to investigate the generic features (if any) of the late stages of crystal growth, using a coarse-grained model of the dynamics[10]. This model is properly viewed as an example of a *cell dynamical system*, discussed in the following section, and is, in some sense, a discrete representation of the *phase field model*, discussed by others at this workshop. The model itself, a complete description of the possible morphologies which may result from diffusion-controlled interface motion, and our resolution of the question of whether or not the generic morphology at large times is that of *diffusion-limited aggregation* or a *dense branching morphology*, are all discussed in detail in ref. 10. Here, we shall just mention an interesting scaling law, observed during our study, and also by other workers using different methods[11,12].

Consider a crystal growing from a planar substrate. We work here in two spatial dimensions, so that the substrate is actually a line. The resulting pattern resembles a complex "forest" of dendrites, growing at more or less equally spaced intervals along the substrate. Denote by $h(x,t)$ the amount of crystal vertically above the point x along the substrate: we shall refer to h as the *height function*. In our model, the variable x is actually discrete, with $1 \leq x \leq L$. The interface width, w, is given by

$$w(t)^2 \equiv \frac{1}{L} \sum_{x=1}^{L} \left(h(x,t) - \bar{h}\right)^2 \tag{11}$$

where $\bar{h} = \sum_x h(x,t)/L$. The power spectrum of h is defined by

$$P(q,t) \equiv \frac{1}{L^2} \left| \sum_{x=1}^{L} h(x,t) \exp(iqx) \right|^2 \tag{12}$$

and its variation with q is found to depend upon the time t. Remarkably, however, the function $P(q,t)/w^2(t)$, is found to be independent of time: when plotted as a function of q for different times t, the data fall onto a single curve.

6. Scaling and universality: Implications for numerical methods. The preceding work is part of a growing body of evidence suggesting that there are universal features shared by a wide class of thermodynamic systems as they approach their equilibrium or asymptotic state. For example, we might be interested in a variable $U(x,t)$ found to obey the self-similar form at long times $U(x,t) = t^{\alpha} f(xt^{\beta})$. Alternatively, U might have wave-like character: $U(x,t) = g(x - vt)$. Without loss of generality, we may consider the self-similar form only, since the substitution $x = \log X$, $t = \log T$ converts the wave-like form into the self-similar form with $\alpha = 0$ and $\beta = -v$. Our goal is to be able to predict the exponents α, β, v... and the scaling functions f, g...

We have found it useful to think of the similarity solutions, corresponding to the (possibly several) intermediate asymptotic states of the system, as fixed points in function space $\mathcal{F}$. The time dependence of the system of interest may then be regarded as the motion of a representative point in $\mathcal{F}$. An intermediate asymptotic regime occurs when the representative point passes near a fixed point. In general, a fixed point will possess an unstable manifold. Thus, although the representative point may be attracted to the fixed point for a finite period of time, in general it will eventually be expelled along the unstable manifold, possibly into the basin of attraction of another fixed point. In the neighbourhood of a fixed point, the dynamics is controlled by the unstable manifold, and is thus a characteristic *solely of the fixed point itself*. Certain types of fixed point may occur generically for a wide class of partial differential equations, possibly describing quite different physical systems. If this is so, then all of the members of this class will exhibit the same intermediate asymptotic behaviour, insofar as the exponents and scaling functions will be the same. This would account for the existence of universal behaviour during the approach to equilibrium, just as the notion of fixed point is believed to account for the universal behaviour observed at a critical point of a thermodynamic system.

To be concrete, consider the familiar example of diffusion. The diffusion equation is usually derived by asserting that the diffusing variable, U, obeys Fick's Law: however, there is nothing in principle (or in practice) which restricts the current to be proportional to ∇U. Thus, in a particular physical situation, a microscopic model might lead to a partial differential equation

$$\partial_t U = a_1 \partial_x^2 U + a_2 \left(\partial_x^2 U\right)^2 + a_3 \left(\partial_x^4 U\right) + \dots \tag{13}$$

where, in this example, a global invariance under $x \to -x$ has been assumed. It may turn out that the long-time behaviour is governed by the fixed point

$$U(x,t) = \frac{1}{\sqrt{t}} G\left(\frac{x}{\sqrt{t}}\right). \tag{14}$$

Then the left hand side and the first term on the right hand side are of $O(t^{-3/2})$ as $t \to \infty$, whereas all other terms are $O(t^{-5/2})$, making suitable assumptions on the class of functions G. Thus, for sufficiently long times, the behaviour of this system will be controlled by the diffusion equation.

This example illustrates a number of salient points. First, from the point of view of determining generic behaviour of a physical system, one should write down all possible terms in the partial differential equation which are consistent with the symmetries of the problem. Second, only some of these terms will be *relevant* at the fixed point of interest: in the example above, these are the terms of $O(t^{-3/2})$ which control the dynamics at long times. Note that there is no reason *a priori* that the relevant terms should be derivable from a Liapunov functional.

Let me now turn to the implications of this viewpoint for numerical methods. If one is only interested in universal behaviour during an intermediate asymptotic regime, then there is no point in solving as precisely as possible the partial differential equation of a particular model of the physical system of interest. In fact, *any* dynamics will suffice, as long as there is reason to believe that it is in the same universality class as the model: if this is so, then the chosen dynamics will eventually reach the neighbourhood of the fixed point corresponding to the intermediate asymptotics of the model. In particular, judiciously chosen dynamics will attain the fixed point much faster and more efficiently than a precise numerical solution of the model.

This observation has been exploited in the *cell dynamical system* approach to spinodal decomposition and crystal growth[13]. There, the physical processes of the system in question are modelled in three steps.

First, the system is divided up into cells whose size is of order the correlation length. Each cell, acting in isolation, obeys a dynamics chosen to mimic the behaviour physically expected. For example, in the Cahn-Hilliard model of spinodal decomposition, the concentration obeys the equation

$$\partial_t c = -\frac{\partial f}{\partial c}. \tag{15}$$

Here, both diffusion and the conserving operator $-\Delta$ have been temporarily ignored. This dynamics has three fixed points corresponding to the initial (unstable) concentration and the two (stable) equilibrium concentrations at coexistence. The dynamics specified by the Cahn-Hilliard model depends upon the parameters τ and u. However, the existence of the fixed point structure is independent of the particular values of τ and u. In the cell dynamical system approach, space and time are discrete. The dynamics of the concentration in cell i at time n, c_i^n are given by

$$c_i^{n+1} = H(c_i^n), \tag{16}$$

where H is chosen to have a similar fixed point structure to eq. (9). For example, $H(x) = A\tanh(Bx)$ is a common choice, with A and B adjustable parameters. Computationally efficient choices of H may be poor representations of the details of the domain wall profile that interpolates between the two stable equilibrium concentrations. This is not a serious disadvantage, since we are interested in the large-scale behaviour of the system, on length scales much greater than the interface width.

Second, the cells are coupled: at time $n+1$, c_i^{n+1} receives a contribution which depends upon the difference between c_i^n and its neighbouring cells. This term mimics diffusion.

Third, global conservation laws are implemented, such as the conservation of mass. The way that this is done is described in detail in ref. 13.

The resulting numerical algorithms are computationally efficient, easily parallelisable, numerically stable, and may be used to investigate the long-time behaviour with relative ease compared to more conventional techniques. In the case of spinodal decomposition and block copolymer phase separation, results from cell dynamical system simulations have been shown to be in agreement with those obtained by more conventional techniques[14,15]. The cell dynamical scheme has been extensively used to study spinodal decomposition in two[13] and three[16] dimensions, and has been used to study the approach to equilibrium of superfluid films (i.e. the XY model of planar ferromagnets)[5], the dynamics of crystal growth[10] and the growth of lamellar phases in two dimensional block copolymer microphase separation[3].

7. Renormalisation group theory for anomalous diffusion. Those readers familiar with the theory of continuous phase transitions will recognise that the picture presented above is similar to renormalisation group (RG) theory[8]. In this section, I elaborate on this connection by presenting an explicit example of a system whose intermediate asymptotics can be solved analytically using the renormalisation group[17,18]. Furthermore, the intermediate asymptotics is by no means trivial: in fact, the exponents cannot be deduced from dimensional analysis at all[1].

The problem in question is the so-called Barenblatt equation[1]

$$\partial_t U = D\partial_x^2 U \tag{17}$$

with $D = 1/2$ for $\partial_x^2 U > 0$ and $D = (1+\epsilon)/2$ for $\partial_x^2 U < 0$. This non-linear diffusion equation describes the pressure U of an elastic fluid in an elasto-plastic porous medium — one which can expand and contract *irreversibly* during the flow process. We are interested in the long time behaviour which results from a bell-shaped initial pressure distribution with a width ℓ. No similarity solution exists of the form $U(x,t) \sim t^{-1/2}G(xt^{-1/2})$.

The renormalisation group method proceeds by posing a solution in the form of a naive perturbation series

$$U = U_0 + \epsilon U_1 + \epsilon^2 U_2 + \ldots \tag{18}$$

By explicit calculation, it is found that

$$U(x,t) = Q\frac{\exp(-x^2/2(t+\ell^2))}{\sqrt{2\pi(t+\ell^2)}}\left\{1 - \frac{\epsilon}{\sqrt{2\pi e}}\ln\left(\frac{t}{\ell^2}\right) + O(\epsilon^2)\right\} + \text{ regular terms.} \tag{19}$$

Here, the regular terms are finite as $t/\ell^2 \to \infty$, and Q is the initial mass of the distribution: $Q = \int U(x,0)\,dx$. The divergence in the perturbation series may be removed by absorbing it into a redefinition of Q. This makes sense because the Barenblatt equation is not a conserving equation. At late times, one cannot deduce the initial mass: it is not an observable.

It can be shown that this renormalisation of Q leads to a renormalised perturbation series

$$U_R(x,t) = Q(t^*)\sqrt{\frac{t^*}{t}}\exp(-x^2/2t)\left\{1 - \frac{\epsilon}{\sqrt{2\pi e}}\ln\left(\frac{t}{t^*}\right) + O(\epsilon^2)\right\}, \tag{20}$$

where $Q(t^*)$ is the value of $U(0,t^*)$, and the subscript R denotes renormalised. This renormalised perturbation series is useful because, given $U(0,t^*)$, the behaviour for $t > t^*$ can be estimated, as long as $(\epsilon/\sqrt{2\pi e})\ln(t/t^*) \ll 1$. For $t \gg t^*$, the expansion is poorly behaved, however. On the other hand, the time t^* is arbitrary: in deriving the renormalised perturbation expansion, no assumption was made about the value of t^*. Clearly, we could in principle adjust t^*, but if we do, we must adjust $Q(t^*)$ by a compensating amount, so that the actual value of the solution $U_R(x,t)$ does not change. Indeed, in the original specification of the problem, no mention was made of t^*, so the solution must be independent of it. i.e.

$$\frac{dU_R}{dt^*} = \frac{\partial U_R}{\partial Q}\frac{dQ}{dt^*} + \frac{\partial U_R}{\partial t^*} = 0. \tag{21}$$

Thus we conclude that

$$t^*\frac{dQ}{dt^*} = -Q\left[\frac{1}{2} + \frac{\epsilon}{\sqrt{2\pi e}} + O(\epsilon^2)\right]. \tag{22}$$

Solving for Q, substituting into the renormalised perturbation expansion and setting $t^* = t$, we obtain

$$U_R(x,t) = \frac{A}{t^{1/2+\alpha}}\exp(-x^2/2t)(1 + O(\epsilon^2)) \tag{23}$$

with

$$\alpha = \frac{\epsilon}{\sqrt{2\pi e}} + O(\epsilon^2) \tag{24}$$

being the *anomalous dimension*. In summary, the renormalisation group has improved the perturbation series, and permitted an asymptotic (at best) expansion of the anomalous dimension.

8. Geometric interpretation of the RG. We can also use the renormalised perturbation expansion to expose the connection between the renormalisation group (as presented in the previous section) and the geometrical picture of fixed points in function space[8]. We define a *renormalisation group transformation* by $U'(x,t_0) \equiv R_b\{U(x,t_0)\}$, where the (in general) non-linear operator R_b performs the following steps: (1) Starting with $U(x,t_0)$, evolve forward in time to $t_1 = bt_0$, $b > 1$. (2) Perform a re-scaling of x: a detailed analysis shows that this must be $x = b^{1/2}x'$. (3) Rescale the amplitude so that $U'(0,t_0) = U(0,t_0)$. Using the renormalised perturbation series to generate the time evolution step, we finally obtain

$$U'(x,t_0) \equiv R_b\{U(x,t_0)\} = Z(b)U(b^{1/2}x, bt_0) \tag{25}$$

where

$$Z(b) = b^{-1/2}(1 - \frac{\epsilon}{\sqrt{2\pi e}}\ln b) + O(\epsilon^2). \tag{26}$$

It is readily verified that to $O(\epsilon)$, the set of transformations R_b for different values of b form a semi-group. A consequence of this result is that $Z(b) = b^{y_Q}$ for some exponent y_Q, i.e.

$$y_Q = \frac{d(\ln Z)}{d(\ln b)}. \tag{27}$$

The RG transformation defines a flow in function space. At the fixed point, $U^* = R_b\{U^*\}$, corresponding to a similarity solution of the original partial differential equation. By choosing $b = 1/t_0$, and performing the differentiation of eqn. (26) to obtain y_Q, we again obtain the result in eqns. (23) and (24). A more complete discussion of the geometric interpretation of the RG is presently in preparation.

This form of the RG lends itself particularly well to numerical computation, and I anticipate that this may become a useful application of the method. In particular, it is hoped to justify and optimise the cell dynamic system approach using this method.

REFERENCES

[1] G.I. Barenblatt, *Similarity, Self-Similarity, and Intermediate Asymptotics*, Consultants Bureau, New York, 1979.

[2] H. Furukawa, *A Dynamic Scaling Assumption for Phase Separation*, Adv. Phys., 34 (1985), pp. 703–750.

[3] M. Bahiana and Y. Oono, *Cell Dynamical System Approach to Block Copolymers*, Phys. Rev. A, 41 (1990), p. 6763.

[4] F. Liu and N. Goldenfeld, *Dynamics of Phase Separation in Block Copolymer Melts*, Phys. Rev. A, 39 (1989), pp. 4805–4810.

[5] M. Mondello and N. Goldenfeld, *Scaling and Vortex Dynamics After the Quench of a System with a Continuous Symmetry*, Phys. Rev. A (to appear).

[6] G. Efstathiou, C. Frenk, S. White and M. Davis, *Gravitational Clustering from Scale-Free Initial Conditions*, Mon. Not. R. astr. Soc., 235 (1988), pp. 715–748.

[7] G.I. Barenblatt, *Dimensional Analysis*, Gordon and Breach Science Publishers, New York, 1987.

[8] S.K. MA, *Modern Theory of Critical Phenomena*, Benjamin/Cummings Publishing Company, Inc, Reading, 1976.

[9] N. GOLDENFELD, *Dynamics of Unstable Interfaces, Physicochemical Hydrodynamics: Interfacial Phenomena*, Plenum Press, New York, 1988, pp. 547–558.

[10] F. LIU AND N. GOLDENFELD, *Generic Features of Late-stage Crystal Growth*, Phys. Rev. A, 42 (1990), pp. 895–903.

[11] R. HARRIS AND M. GRANT, *Monte Carlo Simulation of a Kinetic Ising Model for Dendritic Growth*, Phys. Rev. A (to appear).

[12] D. JASNOW AND J. VINALS, *Dynamical Scaling During Interfacial Growth in the One-Sided Model*, Phys. Rev. A, 40 (1989), p. 3864.

[13] Y. OONO AND A. SHINOZAKI, *Cell Dynamical Systems*, Forma, 4 (1989), pp. 75–102.

[14] E.T. GAWLINSKI, J.D. GUNTON AND J. VINALS, *Domain Growth and Scaling in the Two Dimensional Langevin Model*, Phys. Rev. B, 39 (1989), p. 7266.

[15] A. CHAKRABARTI, R. TORAL AND J.D. GUNTON, *Microphase Separation in Block Copolymers*, Phys. Rev. Lett., 63 (1989), p. 2661.

[16] A. SHINOZAKI AND Y. OONO, *Asymptotic Form-Factor for Spinodal Decomposition in 3-Space*, Phys. Rev. Lett. (to appear).

[17] N. GOLDENFELD, O. MARTIN, Y. OONO AND F. LIU, *Anomalous Dimensions and the Renormalisation Group in a Nonlinear Diffusion Process*, Phys. Rev. Lett., 64 (1990), pp. 1361–1364.

[18] N. GOLDENFELD, O. MARTIN AND Y. OONO, *Intermediate Asymptotics and Renormalisation Group Theory*, J. Sci. Comp., 4 (1989), pp. 355–372.

EVOLVING PHASE BOUNDARIES IN THE PRESENCE OF DEFORMATION AND SURFACE STRESS

MORTON E. GURTIN*

Recent papers of Gurtin [1986, 1988a, 1988b], Angenent and Gurtin [1989], and Gurtin and Struthers [1990] form an investigation whose goal is a *nonequilibrium* thermomechanics of two-phase continua based on Gibb's notion of a sharp phase-interface endowed with energy, entropy and superficial force. In all of these studies except the last the crystal is *rigid*, an assumption that forms the basis of a large class of problems discussed by material scientists, but there are situations in which deformation is the paramount concern, examples being shock-induced transformations and mechanical twinning. In this note I discuss the results of Gurtin and Struthers [1990][1], who consider *deformable* crystal-crystal systems with coherent interface.

One of the chief differences between theories involving phase transitions and the more classical theories of continuum mechanics is the presence of *accretion*, the creation and deletion of material points as the phase interface moves relative to the underlying material, and the interplay between accretion and deformation leads to conceptual difficulties. Three force systems are needed:[2] *deformational forces* to be identified with the forces that act in response to the motion of material points; *accretive forces* that act within the crystal lattice to drive the crystallization process; *attachment forces* associated with the attachment and release of atoms as they are exchanged between phases.

Because of the nonclassical nature of these force systems, it is not at all clear whether there should be additional balance laws, let alone what they should be and how they should relate to the classical momentum balance laws. For that reason *most considerations of this nature are based on invariance*. A new idea, that of *lattice observers*, is introduced: these observers study the crystal lattice and measure the velocity of the accreting crystal surface; they act in addition to the standard *spatial observers*, who measure the gross velocities of the continuum.

The paper is devoted entirely to the physics of the phase interface[3], and for that reason infinitesimally thin control volumes are used; such control volumes

---

*Department of Mathematics, Carnegie Mellon University, Pittsburgh, PA, 15213.

[1] This study was motivated by papers of Cahn [1980], Mullins [1982, 1984], Cahn and Larche [1982], Alexander and Johnson [1985, 1986], and (especially) Leo and Sekerka [1989], all of whom consider deformable media and derive *equilibrium* balance laws for the interface as Euler-Lagrange equations for a global Gibbs function to be stationary.

[2] That more than one force system is needed is clear from a discussion of Cahn [1980], who writes: "solid surfaces can have their physical area changed in two ways, either by creating or destroying surface without changing surface structure and properties per unit area, or by an elastic strain . . . . . along the surface keeping the number of surface lattice sites constant while changing the form, physical area and properties" (cf. Gibbs [1878] pp. 314–331).

[3] The basic equations satisfied by the *bulk material* are the standard equations of a one-phase material and can be found, e.g., in Gurtin [1981].

contain a portion of the interface plus the immediately adjacent bulk material. A basic ingredient of the theory is the *mechanical production* (the outflow of kinetic energy minus the expended power) associated with a control volume. The first law of thermodynamics requires that this production be balanced by the addition of heat and by changes in the internal energy; since heat and energy are invariant quantities, it seems reasonable to presume that the mechanical production itself be invariant. This invariance is used to derive several important results: invariance under changes in the kinetic description of the interface reduces the tangential part of the total accretive stress to a *surface tension*; invariance under changes in spatial and lattice observer yields the mechanical balance laws of the theory. This latter use of invariance is highly nontrivial: it not only leads to the expected momentum balance laws for the surface, it leads to *additional force and moment balance laws for the accretive system*.

It is shown that the power expended on an arbitrary control volume (containing the interface) can be decomposed into: *power expended by surface tension in the creation of new surface, power expended in changing the orientation of the surface, power expended in stretching the surface, power expended by the attachment forces in the exchange of atoms between phases, and inertial power expended in the velocity change between phases*.

The conceptual difficulties of the theory concern forces and the manner in which they relate to the underlying kinematics. For that reason a purely mechanical theory is developed. The underlying thermodynamical law is a dissipation inequality for control volumes: the energy increase plus the energy outflow cannot be greater than the power expended, the relevant energies being the energy of the interface and the bulk energy of the two phases. Again invariance provides an important result: *surface tension equals interfacial energy*.

As constitutive equations the surface energy, the accretive and deformational surface stresses, and the normal attachment force are allowed to depend on the bulk deformation gradient $\mathbf{F}$, the normal $\mathbf{n}$ to the interface, the normal speed v of the interface, and a list $\mathbf{z}$ of subsidiary variables of lesser importance. It follows, as a consequence of the dissipation inequality, that: the surface energy and the accretive and deformational surface stresses are independent of v and $\mathbf{z}$, and depend on $\mathbf{F}$ at most through the tangential deformation gradient $\boldsymbol{F}$; in fact, the energy

$$\psi = \hat{\psi}(\boldsymbol{F}, \mathbf{n}) \tag{1}$$

completely determines the surface stresses through relations, the two most important of which are:

$$\boldsymbol{S} = \partial_{\boldsymbol{F}} \hat{\psi}(\boldsymbol{F}, \mathbf{n}), \qquad \boldsymbol{c} = -D_{\mathbf{n}} \hat{\psi}(\boldsymbol{F}, \mathbf{n}), \tag{2}$$

in which $\boldsymbol{S}$ is the deformational (Piola-Kirchhoff) surface stress, $\boldsymbol{c}$ is the normal accretive stress, $\partial_{\boldsymbol{F}}$ is the partial derivative with respect to $\boldsymbol{F}$, and $D_{\mathbf{n}}$ is the derivative with respect to $\mathbf{n}$ following the interface. A further consequence of the dissipation inequality is an explicit expression for the normal attachment force π:

$$\pi = \mathcal{K} + \Psi + \beta v, \quad \beta = \hat{\beta}(\mathbf{F}, \mathbf{n}, v, \mathbf{z}) \geq 0, \tag{3}$$

where Ψ is the difference in bulk energies, while $\mathcal{K}$ is related to changes in momentum and kinetic energy across the interface. These results imply that the sole source of dissipation is the exchange of atoms between phases, with βv^2 the dissipation per unit interfacial area.

The system of constitutive equations and balance laws combine to give the interface conditions[4]

$$
\begin{aligned}
&\operatorname{div}_{\mathcal{S}} \boldsymbol{S} + (\mathbf{S}_e - \mathbf{S}_c)\mathbf{n} = \rho v(\mathbf{v}_c - \mathbf{v}_3), \\
&\Psi_c - \Psi_e = (\mathbf{S}_c\mathbf{n})\cdot(\mathbf{F}_c\mathbf{n}) - (\mathbf{S}_e\mathbf{n})\cdot(\mathbf{F}_e\mathbf{n}) - \mathcal{K} - \mathcal{G} - \beta v,
\end{aligned} \tag{4}
$$

with

$$
\begin{aligned}
&\mathcal{K} = \frac{1}{2}\rho v^2\{|\mathbf{F}_c\mathbf{n}|^2 - |\mathbf{F}_e\mathbf{n}|^2\}, \\
&\mathcal{G} = \psi\kappa - \operatorname{div}_{\mathcal{S}} \boldsymbol{C} + (\mathbf{F}^T \boldsymbol{S})\cdot \boldsymbol{L}.
\end{aligned} \tag{5}
$$

The subscripts c and e denote the two phases; Ψ_c and Ψ_e are the bulk energies per unit reference volume; $\mathbf{S}_c$ and $\mathbf{S}_e$ are the bulk Piola-Kirchhoff stresses; $\mathbf{F}_c$ and $\mathbf{F}_e$ are the bulk deformation gradients; $\mathbf{v}_c$ and $\mathbf{v}_e$ are the material velocities; ρ is the reference density. The remaining quantities concern the interface: $\boldsymbol{L}$ is the curvature tensor with κ, its trace, the total curvature; $\operatorname{div}_{\mathcal{S}}$ is the surface divergence.

The theory is generalized to include thermal influences. Temperature, bulk and superficial internal energies and entropies, and heat flow are introduced and laws of energy balance and entropy growth are postulated. The constitutive equations are generalized to allow for a dependence on the temperature θ, and an additional constitutive equation for the superficial entropy s is added. In place of (1)-(3), one then has the relations

$$
\begin{aligned}
&\psi = \widehat{\psi}(\boldsymbol{F},\theta,\mathbf{n}), && s = -\partial_\theta\widehat{\psi}(\boldsymbol{F},\theta,\mathbf{n}), \\
&\boldsymbol{S} = \partial_{\boldsymbol{F}}\widehat{\psi}(\boldsymbol{F},\theta,\mathbf{n}), && \boldsymbol{C} = -D_{\mathbf{n}}\widehat{\psi}(\boldsymbol{F},\theta,\mathbf{n}), \\
&\pi = \mathcal{K} + \Psi + \beta v, && \beta = \widehat{\beta}(\mathbf{F},\mathbf{n},\theta,v,\mathbf{z}) \geq 0,
\end{aligned} \tag{6}
$$

with ψ the interfacial free energy.

The resulting interface conditions consist of (4) and (5) in conjunction with the entropy balance[5]

$$
s^\circ - s\kappa v + [S_c - S_e]v = \theta^{-1}[\mathbf{h}_c - \mathbf{h}_e]\cdot\mathbf{n} + \theta^{-1}\beta v^2, \tag{7}
$$

with S_c and S_e the bulk entropies per unit reference volume, $\mathbf{h}_c$ and $\mathbf{h}_e$ the bulk (Piola-Kirchhoff) heat flux vectors per unit reference area, and s° the time-derivative of s following the interface.

Acknowledgement. The research presented here was supported by the Army Research Office and the National Science Foundation.

[4] For statical situations: $(4)_1$ was derived by Gurtin and Murdoch [1975] as a consequence of balance of forces; $(4)_2$ and its counterpart for crystal-melt interactions were derived by Leo and Sekerka [1989] (cf. Johnson and Alexander [1985, 1986]) as Euler-Lagrange equations for stable equilibria. In the absence of surface stress and surface energy ($\boldsymbol{S} = 0$, $\boldsymbol{C} = 0, \psi = 0$): $(4)_1$ is a standard shock relation; $(4)_2$ (with $\beta \neq 0$) was established by Abeyaratne and Knowles [1990]. Counterparts of (4) for a rigid crystal in an inviscid melt were derived by Gurtin [1990]; an analog of $(4)_2$ for a rigid system was given by Gurtin [1988b].

[5] Abeyaratne and Knowles [1990], for the case in which surface stress, surface energy, and surface entropy are neglected.

REFERENCES

[1878] GIBBS, J.W., *On the equilibrium of hetrogeneous substances*, Trans. Connecticut Acad. 3 (1878), pp. 108–248; Reprinted in: *The Scientific Papers of J. Willard Gibbs*, vol. 1, Dover, New York (1961).

[1975] GURTIN, M.E. AND I. MURDOCH, *A continuum theory of elastic material surfaces*, Arch. Rational Mech. Anal., 57, pp. 291–323.

[1980] CAHN, J.W., *Surface stress and the chemical equilibrium of small crystals. 1. The case of the isotropic surface*, Act. Metall. 28, pp. 1333–1338.

[1981] GURTIN, M.E., *An Introduction to Continuum Mechanics*, Academic Press, New York.

[1982] CAHN, J.W. AND F.C. LARCHE, *Surface stress and the chemical equilibrium of small crystals. 2. Solid particles embedded in a solid matrix*, Act. Metall. 30, pp. 51–56.

[1982] MULLINS, W.W., *Thermodynamics of crystal phases with curved interfaces: special case of interface isotropy and hydrostatic pressure*, Proc. Int. Conf. Sol.-Sol. Phase Trans. (Carnegie Mellon) TMS-AIME (eds Aronson, H.I., D.E. Laughlin, R.F. Sekerka, C.M. Wayman) Warrandale, Pa., pp. 49–66.

[1984] MULLINS, W.W., *Thermodynamic equilibrium of a crystalline sphere in a fluid*, J. Chem. Phys., 61, pp. 1436–1442.

[1985] ALEXANDER, J.I.D. AND W.C. JOHNSON, *Thermomechanical equilibrium of solid-fluid systems with curved interfaces*, J. Appl. Phys. 58, pp. 816–824.

[1986] GURTIN, M.E., *On the two-phase Stefan problem with interfacial energy and entropy*, Arch. Rational Mech. Anal. 96, pp. 199–241.

[1986] JOHNSON, W.C. AND J.I.D. ALEXANDER, *Interfacial conditions for thermomechanical equilibrium in two-phase crystals*, J. Appl. Phys. 59, pp. 2735–2746.

[1988a] GURTIN, M.E., *Toward a nonequilibrium thermodynamics of two-phase materials*, Arch. Rational Mech. Anal., 100, pp. 275–312.

[1988b] GURTIN, M.E., *Multiphase thermomechanics with interfacial structure. 1. Heat conduction and the capillary balance law*, Arch. Rational Mech. Anal., 104, pp. 185–221.

[1989] ANGENENT, S. AND M.E. GURTIN, *Multiphase thermomechanics with interfacial structure. 2. Evolution of an isothermal interface*, Arch. Rational Mech. Anal., 108, pp. 323–391.

[1989] LEO, P.H. AND R.F. SEKERKA, *The effect of surface stress on crystal-melt and crystal-crystal equilibrium*, Forthcoming.

[1990] GURTIN, M.E., *A mechanical theory for crystallization of a rigid solid in a liquid melt; melting-freezing waves*, Arch. Rational Mech. Anal., 110, pp. 287–312.

[1990] ABEYARATNE, R. AND J.K. KNOWLES, *On the driving traction acting on a surface of strain discontinuity in a continuum*, J. Mech. Phys. Solids, 38, pp. 345–360.

[1990] GURTIN, M.E. AND A. STRUTHERS, *Multiphase thermomechanics with interfacial structure. 3. Evolving phase boundaries in the presence of bulk deformation*, Arch. Rational Mech. Anal, Forthcoming.

EFFECT OF MODULATED TAYLOR-COUETTE FLOWS ON CRYSTAL-MELT INTERFACES: THEORY AND INITIAL EXPERIMENTS

G. B. MCFADDEN, B. T. MURRAY, S. R. CORIELL*,
M. E. GLICKSMAN AND M. E. SELLECK†

Abstract. An important problem in the process of crystal growth from the melt phase is to understand the interaction of the crystal-melt interface with fluid flow in the melt. This area combines the complexities of the Navier-Stokes equations for fluid flow with the nonlinear behavior of the free boundary representing the crystal-melt interface. Some progress has been made by studying explicit flows that allow a base state corresponding to a one-dimensional crystal-melt interface with solute and/or temperature fields that depend only on the distance from the interface. This allows the strength of the interaction between the flow and the interface to be assessed by a linear stability analysis of the simple base state. The case of a time-periodic Taylor-Couette flow interacting with a cylindrical crystalline interface is currently being investigated both experimentally and theoretically; some preliminary results are given here. The results indicate that the effect of the crystal-melt interface in the two-phase system is to destabilize the system by an order of magnitude relative to the single-phase system with rigid walls.

1. INTRODUCTION

Crystal growth from the liquid or melt phase is often accompanied by hydrodynamic flows in the melt. These flows can have an important impact on the quality of the crystal produced, so that studies of the basic flow-interface interaction are of fundamental importance. In many technologically important techniques, such as Bridgman growth or Czochralski growth, strong electromagnetic fields are frequently used in efforts to control the flow effects. (Some descriptions of common crystal growth techniques are given by Hurle and Jakeman in [1] and references therein.) Avoiding natural convection is one of the main motivations for developing the capability of crystal growth under the microgravity conditions available in low earth orbit, where the driving force for natural convection is lower by orders of magnitude. In addition, such an environment allows more precise fundamental experiments on interface dynamics to be performed without the complicating effects of buoyancy-driven convection [2].

When growing crystals of doped semiconductors or metallic alloys, the concentration of solute at the freezing interface is of special concern [3]. In most applications it is desirable to produce crystals with homogeneous distributions of solute throughout the crystal, and great care is taken in the design of the crystal growth apparatus to attempt to control the concentration and thermal fields near the interface. In directional solidification from the melt, for example, an idealized furnace that is free from imperfections would produce a planar crystal-melt interface with one-dimensional temperature and solute fields, allowing solute to be incorporated uniformly in the growing crystal. In reality, even such a planar crystal-melt interface is subject to various instabilities [4,5] which can result in segregation of solute at the interface and

* National Institute of Standards and Technology, Gaithersburg, MD 20899

† Department of Materials Engineering, Rensselaer Polytechnic Institute, Troy, NY 12180-3590

produce inhomogeneous distributions of solute in the crystal. In addition to instabilities associated with the interface itself [6], under terrestrial growth conditions it is often difficult to avoid the occurrence of fluid flow in the melt due to natural convection [7]. Such flows are themselves able to cause undesirable segregation of solute and may result in the production of inferior quality crystal.

The study of the interaction of fluid flow with a crystal-melt interface is thus an area of fundamental importance in materials science, but despite much recent research [5,8,9] the understanding of such interactions is fragmentary. The general problem combines the complexities of the Navier-Stokes equations for the fluid flow in the melt with the nonlinear behavior of the free boundary representing the crystal-melt interface. Some progress has been made by studying explicit flows that allow a base state corresponding to a one-dimensional crystal-melt interface with solute and/or temperature fields that depend only on the distance from the interface. This allows the strength of the interaction between the flow and the interface to be assessed by a linear stability analysis of the simple base state.

For example, one can examine changes in the *morphological stability* [6] of the interface in the presence of flow in the melt. Specific flows that have been considered in this way include plane Couette flow [10, 11], thermosolutal convection [7,12], plane stagnation flow [13,14], rotating disk flow [15], and the asymptotic suction profile [16]. One can also examine changes in the *hydrodynamic stability* of a given flow that occur when a rigid bounding surface is replaced by a crystal-melt interface. Examples here include the instabilities associated with Rayleigh–Bénard convection [17], thermosolutal convection [8,18,19], plane Poiseuille flow [20], the asymptotic suction profile [16], thermally-driven flow in an annulus [20,21], and Taylor-Couette flow [22,23].

In this paper we consider the influence of the crystal-melt interface on another form of the classical Taylor-Couette instability [24,25,26] of the flow between rotating concentric cylinders. The flow we consider is for a time-dependent sinusoidal rotation of the system, which gives rise to an unsteady, time-periodic base state, whose stability may be assessed by applying Floquet theory [27]. In the next section we give a brief description of some preliminary experimental studies of such a system. We then describe the linear stability problem, and the numerical methods we have implemented for the calculation of neutral stability curves for the system. We close with some numerical results and a comparison with experiment in the final section. In an appendix we describe the implementation of a simple method for determining vorticity boundary values for use in the time-dependent vorticity-streamfunction formulation of the linear governing equations.

2. EXPERIMENT

To examine the interaction between time-periodic Taylor-Couette flow and a crystal melt interface, experiments are currently underway using succinonitrile (SCN), which is a transparent organic material with well-characterized material properties [21]. The melting point of SCN is $T_M = 58$ C, which makes it convenient to use for solidification studies. In addition, unlike many other transparent organic materials, it does not facet upon solidification, and in this and other respects it acts much as a typical metallic material.

To avoid complications with sliding seals that would allow differential rotation in an annular system, it is convenient to seal the SCN within an ampoule with rigidly-connected endcaps and single-unit construction. The entire unit is then oscillated torsionally about the cylindrical axis, so that there is no differential rotation between the inner and outer cylindrical bounding surfaces of the ampoule. A coolant is circulated through the inner cylinder of radius R_0 to maintain a uniform temperature $T_0 < T_M$ at the inner wall. The ampoule is also placed in a heating bath that maintains a uniform temperature $T_2 > T_M$ at the outer cylinder of radius R_2 (see Fig. 1). The SCN filling the region between R_0 and R_2 then melts near the hot outer cylinder and is solid near the cold inner cylinder. The crystal-melt interface separating the two phases, under steady conditions, is found to be cylindrical and co-axial with the annulus, provided that the temperature difference across the melt is small enough. If the temperature difference is held fixed and the ampoule executes sinusoidal rigid-body rotation about the cylindrical axis, then with increasing driving force the cylindrical interface is observed to become unstable and assume an axisymmetric deformed state with a well-defined axial wavelength of the deformation. An example is shown in Fig. 2. Here the view is from the side of the ampoule; the centerline of the annulus lies in the plane of the figure and runs from bottom to the top. The innermost cylinder corresponds to the cool inner wall. Next is the crystal bounded by the deformed axisymmetric crystal-melt interface; two periods of the instability are shown. The interface is surrounded by the melt in the region next to the heated outer cylinder.

The imposed driving force may be characterized by an amplitude and an oscillation frequency, and the stability threshold may be characterized by their marginal values at the onset of instability. For a given frequency of rotation, values of the amplitude at the onset and wavelength can therefore be measured. If the ampoule is long enough, measurements taken near the midsection of the ampoule away from the endcaps are not greatly affected by end conditions, and the results may be interpreted by formulating a linear stability problem for an idealized infinite system.

From the viewpoint of interface stability in the absence of flow, the cylindrical interface would be expected to be stable in this configuration; the stabilizing effect of the radial temperature gradient in the system would be expected to damp out any interface perturbations. On the other hand, the system is subject to hydrodynamic instabilities that can alter significantly the heat transfer in the system. When convective heat transfer to the interface dominates the transport of heat by conduction, the associated thermal variations would be expected to influence the interface shape; the interface would then be expected to reflect the characteristic length scales of the accompanying fluid motion. The change in shape of the crystal-melt interface is also

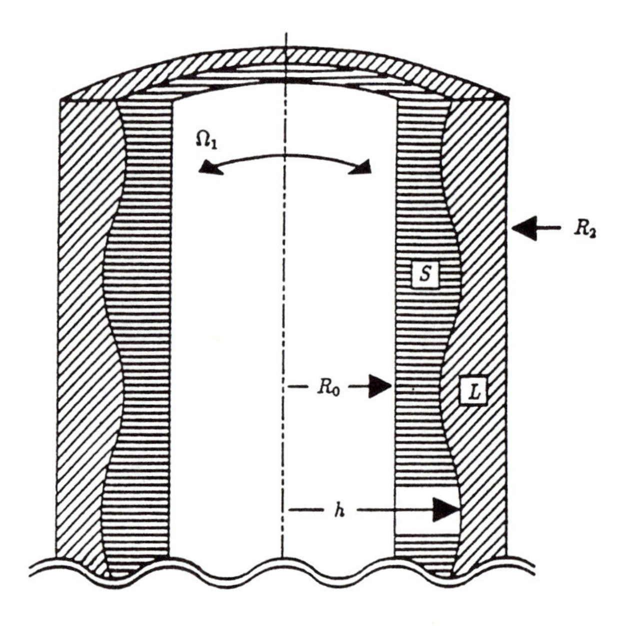

Figure 1: Schematic diagram of the crystalline inner annulus (labeled "S") surrounded by the liquid phase (labeled "L"). In the base state the unperturbed crystal-melt interface is cylindrical, with $h(z,t) = R_1$.

liable to alter the stability characteristics of the flow, and the overall stability of the system in the presence of the crystal-melt interface may differ significantly from the stability of an analogous system with rigid, isothermal boundaries.

By examining the trajectories of small neutrally-dense particles in the liquid it is possible to confirm that the interface deformations are associated with cellular fluid motions that are set up when Taylor-vortices develop in the surrounding fluid. The cellular flow tends to transport hot fluid towards the interface and cold fluid away from the interface, which melts back and freezes in response to the thermal variations. In so doing the stability of the system is found to be diminished by an order of magnitude relative to a rigid-walled system. For steady rotations [22,23], the size of the effect is found to vary with the thermal properties of the melt, and the system is increasingly destabilized as the Prandtl number increases; the Prandtl number $P_r = \nu/\kappa_L$ is the ratio of the kinematic viscosity to the thermal diffusivity.

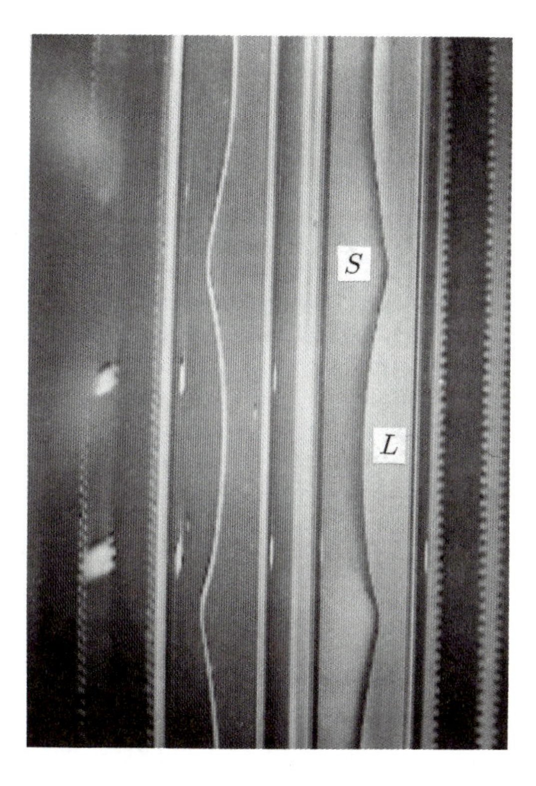

Figure 2: Profile of the instability of a temporally-modulated crystal-melt interface, with a heated outer cylinder and a cooled inner cylinder. The material is succinonitrile, with Prandtl number $P_r = 22.8$. Threaded rod on right provides a scale of 1.27 mm/thread.

3. GOVERNING EQUATIONS

We proceed to consider the idealized case of an annulus of infinite length, and consider the linear stability of a cylindrical crystal-melt interface to axisymmetric perturbations that are sinusoidal in the axial direction with a given axial wavenumber. Buoyancy effects caused by variations of the density with temperature are neglected, and the governing equations are taken to be the incompressible Navier-Stokes equations and the convection-diffusion equation for heat transport.

In cylindrical coordinates the shape of the crystal-melt interface is taken to be $r = h(z,t)$. The liquid occupies the region $h(z,t) < r < R_2$, and the crystal occupies the region $R_0 < r < h(z,t)$. We consider time-dependent rotation of the system; the inner and outer cylinders, and the crystal, all rotate with angular velocity $\Omega_1(t)$.

The axisymmetric nonlinear dimensional governing equations in the liquid take the form

$$\nabla \cdot \mathbf{u} = 0, \tag{1a}$$

$$\frac{\partial \mathbf{u}}{\partial t} + (\mathbf{u} \cdot \nabla)\mathbf{u} + \frac{1}{\rho}\nabla p = \nu \nabla^2 \mathbf{u}, \tag{1b}$$

$$\frac{\partial T_L}{\partial t} + (\mathbf{u} \cdot \nabla) T_L = \kappa_L \nabla^2 T_L, \tag{1c}$$

and in the solid

$$\frac{\partial T_S}{\partial t} = \kappa_S \nabla^2 T_S, \tag{1d}$$

Here the velocity $\mathbf{u}$ has components u, v, and w in the r, ϕ, and z directions, respectively, p is the pressure, ρ is the (constant) density, ν is the kinematic viscosity, T_L and T_S are the temperatures in the liquid and solid, respectively, and κ_L and κ_S are the thermal diffusivities in the liquid and solid, respectively. The solid is rotating with an azimuthal velocity $r\Omega_1(t)$.

The system will be assumed to be periodic in the axial direction with a given wavelength. At the outer boundary $r = R_2$, we have $u = w = 0$, $v = R_2\Omega_1(t)$, and $T_L = T_2$, where T_2 is the (constant) temperature imposed at the outer cylinder; T_2 is assumed to exceed the melting point T_M of the material. At the inner boundary $r = R_0$, we have $T_S = T_0$, where T_0 is the (constant) temperature imposed at the inner cylinder; T_0 is assumed to lie below T_M. At the interface $r = h(z,t)$ the boundary conditions are

$$u = w = 0, \tag{2a}$$

$$v = h(z,t)\Omega_1(t), \tag{2b}$$

$$T_L = T_S = T_M - T_M \Gamma \mathcal{K}, \tag{2c}$$

$$-L_V \frac{\partial h}{\partial t} = k_L \left(\frac{\partial T_L}{\partial r} - \frac{\partial h}{\partial z}\frac{\partial T_L}{\partial z} \right) - k_S \left(\frac{\partial T_S}{\partial r} - \frac{\partial h}{\partial z}\frac{\partial T_S}{\partial z} \right), \tag{2d}$$

where Γ is a capillary length, L_V is the latent heat of fusion per unit volume of crystal, and k_L and k_S are the thermal conductivities in the liquid and solid, respectively. We have assumed equal densities of crystal and melt, and equal heat capacities in each phase.

4. BASE STATE

The imposed rotation is assumed to consist of sinusoidal motion about zero mean, with $\Omega_1(t) = \Omega \cos \omega t$. The base state is given by $h(z,t) = R_1$, $u = v = 0$, $v = v^{(0)}(r,t)$, $p = p^{(0)}(r,t)$, $T_L = T_L^{(0)}(r)$, and $T_S = T_S^{(0)}(r)$. Dimensionless variables may be chosen as follows. The length scale is chosen to be the liquid gap width $L = R_2 - R_1$, the time scale is chosen to be L^2/ν, the velocity scale is chosen to be ν/L, and the deviation of the temperature from its unperturbed value at the interface is measured in units of $\Delta T = T_2 - T_M + T_M\Gamma/R_1$. We retain the same notation for all variables, which will henceforth be dimensionless. The liquid region then occupies the range $\eta/[1-\eta] < r < 1/[1-\eta]$, the unperturbed interface is located at $r = \eta/[1-\eta]$, and the crystal occupies the region $\eta_S/[1-\eta] < r < \eta/[1-\eta]$, where $\eta = R_1/R_2$ and $\eta_S = R_0/R_2 < \eta$. This scaling also introduces the Reynolds number $\mathrm{Re} = L^2\Omega/\nu$, which is a measure of the intensity of the base flow.

The base velocity can be written in the form

$$v^{(0)}(r,t) = \text{Real}\{V(r)\exp i\omega t\},$$

where the complex function $V(r)$ satisfies

$$i\omega V = \left(\frac{\partial^2 V}{\partial r^2} + \frac{1}{r}\frac{\partial V}{\partial r} - \frac{V}{r^2}\right),$$

with $V(1/[1-\eta]) = \text{Re}/[1-\eta]$ and $V(\eta/[1-\eta]) = \eta\text{Re}/[1-\eta]$. The solution to this equation can be expressed in terms of Kelvin functions [28].

The dimensionless temperature fields are given by

$$T_L^{(0)}(r) = \frac{\log(r[1-\eta]/\eta)}{\log(1/\eta)}, \tag{3a}$$

and

$$T_S^{(0)}(r) = \frac{\log(r[1-\eta]/\eta)}{q\log(1/\eta)}, \tag{3b}$$

where $q = k_S/k_L$ is the ratio of thermal conductivities.

5. LINEARIZED EQUATIONS

The linearized governing equations are obtained by writing the variables in the form

$$\begin{pmatrix} u(r,z,t) \\ v(r,z,t) \\ w(r,z,t) \\ p(r,z,t) \\ T_L(r,z,t) \\ T_S(r,z,t) \\ h(z,t) \end{pmatrix} = \begin{pmatrix} 0 \\ v^{(0)}(r,t) \\ 0 \\ p^{(0)}(r,t) \\ T_L^{(0)}(r) \\ T_S^{(0)}(r,t) \\ \eta/(1-\eta) \end{pmatrix} + \begin{pmatrix} u^{(1)}(r,t) \\ v^{(1)}(r,t) \\ w^{(1)}(r,t) \\ p^{(1)}(r,t) \\ T_L^{(1)}(r,t) \\ T_S^{(1)}(r,t) \\ h^{(1)}(t) \end{pmatrix} \exp(iaz),$$

where the latter perturbations quantities are assumed to be small. These expressions are substituted into the nonlinear equations, and the equations and boundary conditions are expanded to first order in the perturbed quantities.

We employ two different numerical approaches for computing the linear stability of the system. The first is implemented for the linearized equations expressed in terms of the velocity and pressure, and the second employs a vorticity-streamfunction formulation.

5.1. Velocity-pressure formulation. The linearized equations in the melt region $\eta/(1-\eta) < r < 1/(1-\eta)$ take the form

$$\frac{\partial u^{(1)}}{\partial r} + \frac{1}{r}u^{(1)} + iaw^{(1)} = 0, \tag{4a}$$

$$\frac{\partial u^{(1)}}{\partial t} + \frac{\partial p^{(1)}}{\partial r} = \frac{\partial^2 u^{(1)}}{\partial r^2} + \frac{1}{r}\frac{\partial u^{(1)}}{\partial r} - \left[a^2 + \frac{1}{r^2}\right]u^{(1)} + 2\frac{v^{(0)}}{r}v^{(1)}, \tag{4b}$$

$$\frac{\partial v^{(1)}}{\partial t} + u^{(1)}\frac{\partial v^{(0)}}{\partial r} = \frac{\partial^2 v^{(1)}}{\partial r^2} + \frac{1}{r}\frac{\partial v^{(1)}}{\partial r} - \left[a^2 + \frac{1}{r^2}\right]v^{(1)} - \frac{v^{(0)}}{r}u^{(1)}, \tag{4c}$$

$$\frac{\partial w^{(1)}}{\partial t} + iap^{(1)} = \frac{\partial^2 w^{(1)}}{\partial r^2} + \frac{1}{r}\frac{\partial w^{(1)}}{\partial r} - a^2 w^{(1)}, \tag{4d}$$

$$\frac{\partial T_L^{(1)}}{\partial t} + u^{(1)}\frac{\partial T_L^{(0)}}{\partial r} = \frac{1}{P_r}\left(\frac{\partial^2 T_L^{(1)}}{\partial r^2} + \frac{1}{r}\frac{\partial T_L^{(1)}}{\partial r} - a^2 T_L^{(1)}\right), \tag{4e}$$

and, in the region $\eta_S/(1-\eta) < r < \eta/(1-\eta)$, one obtains

$$\frac{\partial T_S^{(1)}}{\partial t} = \frac{1}{P_s}\left(\frac{\partial^2 T_S^{(1)}}{\partial r^2} + \frac{1}{r}\frac{\partial T_S^{(1)}}{\partial r} - a^2 T_S^{(1)}\right). \tag{4f}$$

Here $P_r = \nu/\kappa_L$ is the Prandtl number and $P_s = \nu/\kappa_S$.

The linearized boundary conditions are $u^{(1)} = v^{(1)} = w^{(1)} = T_L^{(1)} = 0$ at $r = 1/(1-\eta)$, $T_S^{(1)} = 0$ at $r = \eta_S/(1-\eta)$, and, at $r_I = \eta/(1-\eta)$,

$$u^{(1)} = w^{(1)} = 0, \tag{5a}$$

$$v^{(1)} = \left(\mathrm{Re}\cos\omega t - \frac{\partial v^{(0)}}{\partial r}\right) h^{(1)}, \tag{5b}$$

$$T_L^{(1)} + \frac{\partial T_L^{(0)}}{\partial r} h^{(1)} = T_S^{(1)} + \frac{\partial T_S^{(0)}}{\partial r} h^{(1)} = -\gamma\left(a^2 - \frac{1}{r_I^2}\right) h^{(1)}, \tag{5c}$$

$$-\mathcal{L}\frac{\partial h^{(1)}}{\partial t} = \left(\frac{\partial T_L^{(1)}}{\partial r} - q\frac{\partial T_S^{(1)}}{\partial r}\right). \tag{5d}$$

where $\gamma = (T_M\Gamma)/(L\Delta T)$ and $\mathcal{L} = (\nu L_V)/(k_L\Delta T)$.

5.2. Vorticity-streamfunction formulation. As an alternative formulation we introduce a Stokes stream function, $\psi^{(1)}(r)\exp(iaz)$, defined so that

$$ru^{(1)} = -ia\psi^{(1)}, \qquad rw^{(1)} = \frac{\partial \psi^{(1)}}{\partial r}; \tag{6}$$

the continuity equation Eq. (4a) is then satisfied identically. The curl of the velocity defines the vorticity vector, whose azimuthal component, $\zeta^{(1)}(r)\exp(iaz)$, satisfies

$$\zeta^{(1)} = -\left\{\frac{\partial}{\partial r}\left(\frac{1}{r}\frac{\partial \psi^{(1)}}{\partial r}\right) - \frac{a^2}{r}\psi^{(1)}\right\}. \tag{7a}$$

The vorticity transport equation is obtained by taking the curl of the momentum equation; the azimuthal component of this equation takes the form

$$\frac{\partial \zeta^{(1)}}{\partial t} - 2ia\frac{v^{(0)}}{r}v^{(1)} = \left(\frac{\partial^2 \zeta^{(1)}}{\partial r^2} + \frac{1}{r}\frac{\partial \zeta^{(1)}}{\partial r} - \left[a^2 + \frac{1}{r^2}\right]\zeta^{(1)}\right). \tag{7b}$$

With this formulation, Eqns. (4a), (4b), and (4d) are replaced by Eqns. (7a) and (7b), and the boundary conditions $u^{(1)} = w^{(1)} = 0$ are replaced by $\psi^{(1)} = \partial\psi^{(1)}/\partial r = 0$; the remaining boundary conditions are unchanged.

6. NUMERICAL TREATMENT

We employ two numerical treatments for this time-dependent linear stability problem. The first, for the velocity-pressure formulation, consists of a Fourier representation in time based on Floquet theory. This produces a large set of coupled two-point boundary values problems in the radius variable that are treated using numerical ODE software. The second approach, for the vorticity-streamfunction formulation, consists of using a Chebyshev pseudospectral method to discretize the radial variable, followed by a numerical integration of the resulting system of coupled ODE's forward in time to monitor growth or decay from cycle to cycle.

The latter approach is very useful for performing fast rough calculations to determine the lowest mode and provide starting guesses for the first procedure, which is more time-consuming but very accurate provided a good starting guess is available. The second approach could also be used to perform a conventional Floquet treatment by determining the eigenvalues of a computed fundamental solution matrix, but we have found it to be simpler just to integrate the initial value problem instead.

6.1. Velocity-pressure treatment. Hall [29] and Seminara and Hall [30] studied a similar time-dependent stability problem by using Floquet theory with a Fourier development in the time variable (see also [31,32]). In this approach, the perturbation quantities are expressed in the form

$$\begin{pmatrix} u^{(1)}(r,t) \\ v^{(1)}(r,t) \\ w^{(1)}(r,t) \\ p^{(1)}(r,t) \\ T_L^{(1)}(r,t) \\ T_S^{(1)}(r,t) \\ h^{(1)}(t) \end{pmatrix} = \exp(\sigma t) \sum_{m=-M}^{M} \begin{pmatrix} u_m(r) \\ v_m(r) \\ w_m(r) \\ p_m(r) \\ T_{Lm}(r) \\ T_{Sm}(r) \\ h_m \end{pmatrix} \exp(im\omega t).$$

The linear governing equations then reduce to the form

$$\frac{\partial u_m}{\partial r} + \frac{1}{r}u_m + iaw_m = 0, \tag{8a}$$

$$(\sigma + im\omega)u_m + \frac{\partial p_m}{\partial r} = \frac{\partial^2 u_m}{\partial r^2} + \frac{1}{r}\frac{\partial u_m}{\partial r} - \left[a^2 + \frac{1}{r^2}\right]u_m + \frac{1}{r}\left\{\bar{V}v_{m+1} + Vv_{m-1}\right\}, \tag{8b}$$

$$(\sigma + im\omega)v_m + \frac{1}{2}\left\{u_{m+1}\frac{\partial \bar{V}}{\partial r} + u_{m-1}\frac{\partial V}{\partial r}\right\} = \frac{\partial^2 v_m}{\partial r^2} + \frac{1}{r}\frac{\partial v_m}{\partial r} - \left[a^2 + \frac{1}{r^2}\right]v_m - \frac{1}{2r}\left\{\bar{V}u_{m+1} + Vu_{m-1}\right\}, \tag{8c}$$

$$(\sigma + im\omega)w_m + iap_m = \frac{\partial^2 w_m}{\partial r^2} + \frac{1}{r}\frac{\partial w_m}{\partial r} - a^2 w_m, \tag{8d}$$

$$(\sigma + im\omega)T_{Lm} + u_m\frac{\partial T_L^{(0)}}{\partial r} = \frac{1}{P_r}\left(\frac{\partial^2 T_{Lm}}{\partial r^2} + \frac{1}{r}\frac{\partial T_{Lm}}{\partial r} - a^2 T_{Lm}\right), \tag{8e}$$

and, in the region $\eta_S/(1-\eta) < r < \eta/(1-\eta)$, one obtains

(8f)
$$(\sigma + im\omega)T_{Sm} = \frac{1}{P_s}\left(\frac{\partial^2 T_{Sm}}{\partial r^2} + \frac{1}{r}\frac{\partial T_{Sm}}{\partial r} - a^2 T_{Sm}\right).$$

for $-M < m < M$; here quantities with subscripts $\pm(M+1)$ are assumed to vanish, and the overbars denote the complex conjugate. Note that because of the appearance of the factor
$$v^{(0)}(r,t) = \frac{1}{2}\left\{V(r)\exp(i\omega t) + \bar{V}(r)\exp(-i\omega t)\right\},$$
there is "tridiagonal" coupling between the neighboring coefficients of index $m-1$, m, and $m+1$.

The linearized boundary conditions are $u_m = v_m = w_m = T_{Lm} = 0$ at $r = 1/(1-\eta)$, $T_{Sm} = 0$ at $r = \eta_S/(1-\eta)$, and, at $r_I = \eta/(1-\eta)$,

(9a)
$$u_m = w_m = 0,$$

(9b)
$$v_m = \frac{1}{2}\left(\mathrm{Re} - \frac{\partial \bar{V}}{\partial r}\right) h_{m+1} + \frac{1}{2}\left(\mathrm{Re} - \frac{\partial V}{\partial r}\right) h_{m-1},$$

(9c)
$$T_{Lm} + \frac{\partial T_L^{(0)}}{\partial r} h_m = T_{Sm} + \frac{\partial T_S^{(0)}}{\partial r} h_m = -\gamma\left(a^2 - \frac{1}{r_I^2}\right) h_m,$$

(9d)
$$-\mathcal{L}(\sigma + im\omega)h_m = \left(\frac{\partial T_{Lm}}{\partial r} - q\frac{\partial T_{Sm}}{\partial r}\right).$$

The interface deformation h_m can be eliminated in Eqns. (9b)–(9d), so that the interface boundary conditions contain only the solution variables that appear in the governing equations (8). Because of Eqn. (9b), coupling is obtained between the $m-1$, m, and $m+1$ modes of the solution variables at the interface.

The order of the equations may be reduced by using the derivative of the continuity equations to eliminate the highest derivative in the radial momentum equation. The equations are then written as a system of first order equations, and solved numerically using the SUPORT subroutines [34] from the SLATEC Common Math Library [35]. SUPORT solves nonsingular two-point boundary values problems by superposition of numerically integrated solutions, with an orthonormalization technique used to assure linear independence of the intermediate solutions. To use SUPORT to solve the homogeneous eigenvalue problem, we implement a method described by H. Keller [36] and replace one homogeneous boundary condition by an inhomogeneous boundary condition for which the system is nonsingular. The original boundary condition is then used as a nonlinear equation for the eigenvalue; the root to this equation is found using the routine SNSQ [37]. To determine neutral modes, it is convenient to set $\sigma_r = 0$ and use the root finder to solve directly for the critical Reynolds number Re.

The stability of the system [33] is governed by the real part σ_r of the complex growth rate σ; the base state is stable if, for all wavenumbers a, the corresponding eigenvalues σ_r are negative, and states of marginal stability correspond to critical wavenumbers a_c for which $\sigma_r = 0$ while $\sigma_r < 0$ for all other wavenumbers. The values of σ_i, the imaginary part of σ, determine the nature of the response to the forcing. If

$\sigma_i = 0$ at the onset, the response is synchronous with the forcing. Another possibility is for the response to have $\sigma_i = \omega/2$, which is a subharmonic response; in this case the period of the response is twice that of the forcing. Other values of σ_i are also possible.

Since we may take the real part of the perturbed quantities to determine the marginal state, it is possible to use symmetry to significantly reduce the number of unknowns in the linear system. The specific nature of the symmetry depends on the value of σ_i. For synchronous response ($\sigma_i = 0$) and for subharmonic response ($\sigma_i = \omega/2$), quantities with negative indices may be eliminated by using the relations

$$\begin{pmatrix} u_{(-m)} \\ v_{(-m)} \\ w_{(-m)} \\ p_{(-m)} \\ T_{L(-m)} \\ T_{S(-m)} \\ h_{(-m)} \end{pmatrix} = \begin{pmatrix} \bar{u}_m \\ \bar{v}_m \\ -\bar{w}_m \\ \bar{p}_m \\ \bar{T}_{Lm} \\ \bar{T}_{Sm} \\ \bar{h}_m \end{pmatrix}, \quad \text{and} \quad \begin{pmatrix} u_{(-m)} \\ v_{(-m)} \\ w_{(-m)} \\ p_{(-m)} \\ T_{L(-m)} \\ T_{S(-m)} \\ h_{(-m)} \end{pmatrix} = \begin{pmatrix} \bar{u}_{(m-1)} \\ \bar{v}_{(m-1)} \\ -\bar{w}_{(m-1)} \\ \bar{p}_{(m-1)} \\ \bar{T}_{L(m-1)} \\ \bar{T}_{S(m-1)} \\ \bar{h}_{(m-1)} \end{pmatrix},$$

respectively. The modes we compute in this work are all synchronous. There is a further symmetry that we have exploited in this case, which gives rise to two classes of synchronous solutions. For the first class, denoted by Type I, the azimuthal velocity components v_m are nonzero only for odd values of m, and the other components u_m, w_m, p_m T_{Lm}, and T_{Sm} are nonzero for even values of m. For the second class (Type II), the reverse is true: v_m are nonzero only for even values of m, and the others are nonzero for odd values of m. The Type I modes are found using the boundary condition $dv_1/dr = 1$ in the implementation of Keller's method, and the Type II modes are found using the condition $dv_0/dr = 1$. Using this symmetry reduces the number of unknowns by another factor of two; the total order of the system of first order equations in r for the liquid region is then $8M + 8$.

6.2. Vorticity-streamfunction treatment. We have also constructed a second numerical procedure to integrate the vorticity-streamfunction formulation of the problem as an initial value problem. We discretize the spatial derivatives using the standard Chebyshev pseudospectral method (see, e.g., [39,40]). For this method it is convenient to rescale the radial variable. In the liquid we set $r = \eta/(1-\eta)+\frac{1}{2}(\xi+1)$ for $-1 < \xi < 1$, and in the solid we set $r = \eta_S/(1-\eta) + \frac{1}{2}(\xi^S + 1)(\eta - \eta_S)/[1-\eta]$ for $-1 < \xi^S < 1$. Thus in the liquid we have $d/dr = 2d/d\xi$, and in the solid we have $d/dr = 2([1-\eta]/[\eta - \eta_S])d/d\xi^S$.

In the liquid we use the points $\xi_j = \cos j\pi/N$ for $j = 0, 1, \ldots, N$, and in the solid we use the points $\xi_j^S = \cos j\pi/N$ for $j = 0, 1, \ldots, N$, so that the outer cylinder corresponds to the point $\xi_0 = 1$, the interface corresponds to the point $\xi_N = -1$ or $\xi_0^S = 1$, and the inner cylinder corresponds to the point $\xi_0^S = -1$. At these collocation points, we use the Chebyshev derivative matrix D [40], which has the property that at the points ξ_j, the derivative g_j' of any N-th degree polynomial $g(\xi)$ is given exactly in terms of its collocation values g_k by the expression

$$g'_j = \sum_{k=0}^{N} D_{jk} g_k.$$

Higher derivatives are represented by powers $D^{(n)}$ of the matrix D.

At the interior points ξ_j of the liquid region, $j = 1, \ldots, N-1$, we have the discrete equations

$$4\sum_{k=0}^{N} D^{(2)}_{jk}\psi^{(1)}_k - \frac{2}{r_j}\sum_{k=0}^{N} D_{jk}\psi^{(1)}_k - a^2\psi^{(1)}_j = -r_j\zeta^{(1)}_j, \tag{10a}$$

$$\frac{\partial\zeta^{(1)}_j}{\partial t} = 4\sum_{k=0}^{N} D^{(2)}_{jk}\zeta^{(1)}_k + \frac{2}{r_j}\sum_{k=0}^{N} D_{jk}\zeta^{(1)}_k - \left[a^2 + \frac{1}{r_j^2}\right]\zeta^{(1)}_j + 2ia\frac{v^{(0)}_j}{r_j}v^{(1)}_j, \tag{10b}$$

$$\begin{aligned}\frac{\partial v^{(1)}_j}{\partial t} - 2ia\frac{\psi^{(1)}_j}{r}\sum_{k=0}^{N} D_{jk}v^{(0)}_k = 4\sum_{k=0}^{N} D^{(2)}_{jk}v_k + \frac{2}{r_j}\sum_{k=0}^{N} D_{jk}v^{(1)}_k \\ - \left[a^2 + \frac{1}{r_j^2}\right]v^{(1)}_j + ia\frac{v^{(0)}_j}{r_j}\psi^{(1)}_j,\end{aligned} \tag{10c}$$

$$\begin{aligned}\frac{\partial T^{(1)}_{Lj}}{\partial t} - 2ia\frac{\psi^{(1)}_j}{r}\sum_{k=0}^{N} D_{jk}T^{(0)}_{Lk} = \frac{1}{P_r}\left(4\sum_{k=0}^{N} D^{(2)}_{jk}T^{(1)}_{Lk}\right. \\ \left. + \frac{2}{r_j}\sum_{k=0}^{N} D_{jk}T^{(1)}_{Lk} - a^2T^{(1)}_{Lj}\right),\end{aligned} \tag{10d}$$

and similarly at the interior points ξ^S_j of the solid region, for $j = 1, \ldots, N-1$ we have

$$\frac{\partial T^{(1)}_{Sj}}{\partial t} = \frac{1}{P_s}\left(4\frac{[1-\eta]^2}{[\eta-\eta_S]^2}\sum_{k=0}^{N} D^{(2)}_{jk}T^{(1)}_{Sk} + \frac{2}{r_j}\frac{[1-\eta]}{[\eta-\eta_S]}\sum_{k=0}^{N} D_{jk}T^{(1)}_{Sk} - a^2T^{(1)}_{Sj}\right), \tag{10e}$$

At the interface we have an ODE for the interface deformation,

$$\frac{\partial h^{(1)}}{\partial t} = \frac{-1}{\mathcal{L}}\left\{\sum_{k=0}^{N}\left(D_{Nk}T^{(1)}_{Lk} - qD_{0k}T^{(1)}_{Sk}\right)\right\}. \tag{10f}$$

We use the ODE's for $h^{(1)}$ and the $4N-4$ interior values $\zeta^{(1)}_j$, $v^{(1)}_j$, $T^{(1)}_{Lj}$, and $T^{(1)}_{Sj}$ for $j = 1, \ldots, N-1$, to advance the solution in time. This can be done provided that the necessary boundary values at $j = 0$ and $j = N$ that are required by the right hand sides of the ODE's can be determined from the boundary conditions, and provided that a solution to the second order two-point boundary value problem (10a) for $\psi^{(1)}_j$ can be obtained that is consistent with the given boundary data for $\psi^{(1)}$.

Given $h^{(1)}$ and the other interior values, boundary values $j = 0$ and $j = N$ for $v^{(1)}_j$, $T^{(1)}_{Lj}$, and $T^{(1)}_{Sj}$ follow immediately from the given boundary conditions. It is not immediately obvious how boundary values for the vorticity may be determined, or how to solve the overdetermined problem for $\psi^{(1)}$. In the Appendix we give a simple algorithm for the determination of consistent vorticity boundary data in terms of the given interior values, and show that the resulting vorticity distribution is such that the solution to the Dirichlet problem for the streamfunction satisfies the Neumann conditions as well.

The procedure to determine growth rates for synchronous modes can therefore be carried out as follows. At time t, given the $4N-3$ values $h^{(1)}$ and $\zeta_j^{(1)}$, $v_j^{(1)}$, $T_{Lj}^{(1)}$, and $T_{Sj}^{(1)}$ for $j = 1, \ldots, N-1$, the vorticity boundary values $\zeta_0^{(1)}$ and $\zeta_N^{(1)}$ are determined by the method outlined in the appendix. Eqn. (10a), with boundary conditions $\psi_0^{(1)} = \psi_N^{(1)} = 0$ is solved using the LINPACK routines SGECO and SGESL [41]; the matrix factorization may be stored as a preprocessing step, and the solution is obtained by backsolving. All the values necessary to determine the time derivatives of the $4N-3$ data points at time t are thus available, and the solution may be advanced in time using a standard ODE package; we use the routine SDRIVE [42].

This scheme is found to be both stable and efficient in this setting. Values of the derivatives of the streamfunction at the endpoints are monitored as functions of time as an error check, and are found to remain small for integration over dozens of periods of the motion. To estimate growth rates, we integrate over a few periods $T = 2\pi/\omega$ of the forcing while monitoring the ratios of successive quantities such as $\zeta_j^{(1)}(t)$, evaluated at fixed points in space, for $t = T$, $2T$, etc. As in the power method for eigenvalue determination [43], after a few cycles a dominant mode emerges, and such ratios tend to expressions of the form $\exp(\sigma_r T)$, as follows from the Floquet representation of the solution. The number of cycles required depends on the relative separation of the dominant and subdominant modes; for this application convergence is usually achieved in ten or twenty cycles. The spatial form of the final solution approximates the dominant mode as well, and can be used as initial data in subsequent calculations when doing continuation. Note that only the most dangerous mode can be determined by this rudimentary procedure; on the other hand, it is useful to know that one is not tracking a higher mode that is not the most dangerous, as can happen with other procedures. Note also that the simple procedure of taking ratios at successive periods to determine σ works well for synchronous or subharmonic modes, but would lead to confusing results for modes with arbitrary values of σ_i.

On a coarse mesh of 12 to 16 points, values for σ_r can be determined quite rapidly, and good approximations to conditions of marginal stability can be determined in a few runs. One can then use the same method with a refined mesh to improve the answer, or use the approximation as a starting guess for the other numerical procedure.

7. RESULTS

We present some numerical calculations for the preliminary experimental work alluded to above. We take $R_2 = 1.60$ cm, $R_1 = 1.11$ cm, and $R_0 = 0.458$ cm, giving $\eta = 0.69$ and $\eta_S = 0.286$. For SCN, the thermal properties of the liquid and solid are about equal, and we take $P_r = 22.8$, $P_s = 22.8$, and $q = 1.0$. The solutions we compute correspond to a *steady* axisymmetric crystal-melt interface; the only non-vanishing component of the interface perturbation $h^{(1)}(t)$ is the mean component ($m = 0$). The results are therefore independent of the specific values used for the dimensionless latent heat $\mathcal{L}$. Since the wavelengths of the instability are large compared to the capillary length, the parameter γ is negligible as well.

Critical values of the Reynolds number Re are shown in Fig. 3 as a function of the dimensionless driving frequency ω. Numerical values are also given in Table I, along with values for the critical axial wavenumber a. For comparison, critical values for Re in the case of a rigid-walled system with $\eta = 0.69$ are given in Table II. For $\omega = 28.9$, for example, the critical Reynolds number Re = 14.9 is an order of magnitude smaller than that for the rigid-walled system, Re = 145.2. The wavelength of the most dangerous disturbance is also longer by a factor of 2.7 for the two-phase system.

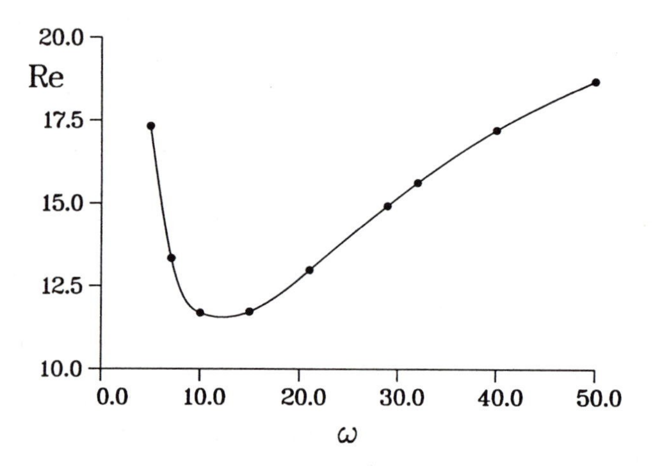

Figure 3: Critical Reynolds numbers Re versus the dimensionless forcing frequency ω for $\eta = 0.69$ and $P_r = 22.8$; values for other paramters are given in the text.

Table I. Marginal Stability with Crystal-Melt Interface

ω	Re	a	Type
5.00	17.3185	1.241	I
7.07	13.3147	1.387	I
10.0	11.6945	1.502	I
15.0	11.7284	1.637	I
21.0	12.9839	1.757	I
28.9	14.9219	1.866	I
32.0	15.6263	1.902	I
40.0	17.1926	1.978	I
50.0	18.6824	2.043	I

Table II. Marginal Stability with Rigid Sidewalls

ω	Re	a	Type
10.0	142.046	5.309	II
12.0	132.082	5.006	I
12.5	130.472	5.069	I
13.0	130.178	5.124	I
15.0	138.319	5.560	I
21.0	133.430	4.800	II
28.9	145.203	5.059	II
32.0	157.692	5.339	II
35.0	177.733	5.519	II

Figure 4 shows values for the the two-phase system (solid curve) and the rigid-wall system (dashed curves). The destabilization occurs over the full range of frequencies studied. The results for the rigid-walled system contain solutions with both Type I and Type II symmetries; the two-phase solutions are all of Type I. We have also included in this figure some preliminary experimental values. Here states that were determined to be marginally unstable are indicated by open circles, and stable states are indicated by solid dots. All data were obtained with an interfacial temperature gradient of +2.6 K/cm. At a frequency of $\omega = 29.6$, a series of unstable states were investigated, the lowest state having a Reynolds number Re $= 21.4$. At a frequency of $\omega = 7.07$, a Reynolds number of Re $= 40.2$ was found to be unstable, and a Reynolds number Re $= 17.4$ was found to be stable. This stable value actually exceeds the numerically-determined critical Reynolds number of Re $= 13.3$. Given the rough nature of the experiments, however, the agreement is reasonable; in particular, the predicted destabilization of the two-phase system relative to the rigid-walled system seems to be borne out.

A spatial error tolerance of 10^{-10} was used in the Adam's-type procedure in SUPORT. Lower-frequency solutions are more complicated temporally, and require a larger number of Fourier modes in time (for example, see [31]). The number of modes required also depends on the relative magnitude of Re; calculations for the rigid-walled system require more modes than those for the two-phase system to achieve comparable accuracy. The rigid-walled results were obtained using $M = 32$ Fourier modes; the last few modes then are on the order of the machine roundoff error. Similar accuracy was obtained for solutions to the two-phase system; here, however, four to sixteen modes sufficed. The numerical results were checked using both numerical procedures.

To obtain quantitative agreement between theory and experiment in future work, several additional factors can be taken into account. For a system aligned vertically with the gravitational field, both axial and radial buoyancy effects could be included

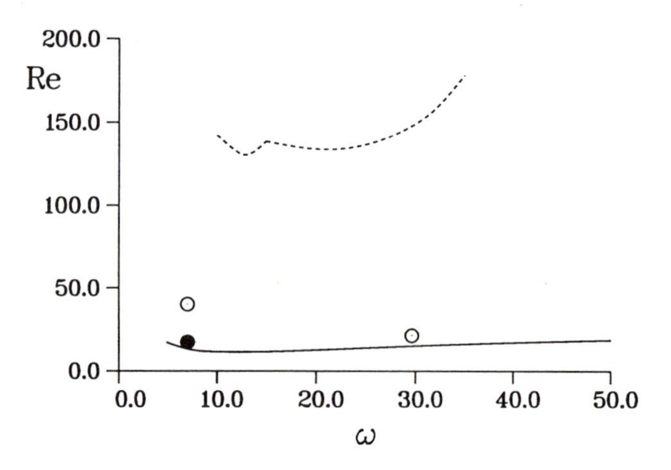

Figure 4: Critical Reynolds numbers Re versus the dimensionless forcing frequency ω. The solid curve is identical to that shown in Fig. 3, and corresponds to the two-phase system. The dashed curves correspond to a rigid-walled system. The data points correspond to experimental values: the open circles indicate marginally unstable states, and the solid circle indicates the stable state.

in the theoretical treatment. In a related study [44], such effects were considered in the context of steady-state forcing. Under conditions similar to those considered here, the interaction of the density gradient with the centripetal acceleration was found to be unimportant, and the gravitational field caused a modest shift in Re and a two-fold increase in wavelength at the onset of instability. The inclusion of a vertical buoyancy force also induces an axial drift of the disturbances, with an associated nonzero phase speed ($\sigma_i \neq 0$). If the gravitational force becomes strong enough, the possibility of non-axisymmetric disturbances must be considered as well [21,45].

ACKNOWLEDGMENTS

The authors are grateful for helpful conversations with R. F. Boisvert, R. Fosdick, and B. V. Saunders. This work was conducted with the support of the Microgravity Science and Applications Division of the National Aeronautics and Space Administration, and the Applied and Computational Mathematics Program of the Defense Advanced Research Projects Agency. One of the authors (BTM) was supported by an NRC Postdoctoral Research Fellowship. Some of the work was performed while one of the authors (GBM) was visiting the Institute for Mathematics and its Applications at the University of Minnesota, whose hospitality is gratefully acknowledged. Numerical calculations were performed at the Minnesota Supercomputer Center through an NSF grant to the IMA.

APPENDIX

Here we describe a simple method for the determination of vorticity boundary values for use in the time integration of the linear governing equations. The technique is based on the observation that the equations for the streamfunction are, in a sense, overdetermined, and the vorticity must therefore obey compatibility conditions for a solution to exist. Our treatment is an elementary application of ideas that have recently been developed in the study of vortex dynamics [46,47].

Suppose $\psi(r)$ satisfies

$$L\psi = \frac{\partial}{\partial r}\left(\frac{1}{r}\frac{\partial \psi}{\partial r}\right) - \frac{a^2}{r}\psi = -\zeta,$$

for $0 < a < r < b$, with $\psi(a) = \psi_r(a) = \psi(b) = \psi_r(b) = 0$. Then if $v^{(p)}(r)$, $p = 1, 2$, is either of two linearly independent solutions to $Lv = 0$, integration by parts shows that

$$\int_a^b v^{(p)}(\xi)\,\zeta(\xi)\,d\xi = 0,$$

so that there are two independent integral constraints on the function ζ in order for a solution $\psi(r)$ to exist. Conversely, construct the Greens function $G(r,\xi)$ for the Dirichlet problem for L from the two solutions $y^{(1)}(r)$ and $y^{(2)}(r)$ to $Ly = 0$ with $y^{(1)}(a) = 0$, $y^{(1)}(b) = 1$, $y^{(2)}(b) = 0$, and $y^{(2)}_r(b) = 1$. The Wronskian $W(r) = y^{(1)}y^{(2)}_r - y^{(1)}_r y^{(2)}$ is then just $W(r) = r$, and we have the simple form

$$G(r,\xi) = \begin{cases} y^{(1)}(r)\,y^{(2)}(\xi), & \text{for } r < \xi \\ y^{(1)}(\xi)\,y^{(2)}(r), & \text{for } \xi < r \end{cases}$$

The representation

$$\psi(r) = -y^{(2)}(r)\int_a^r y^{(1)}(\xi)\,\zeta(\xi)\,d\xi - y^{(1)}(r)\int_r^b y^{(2)}(\xi)\,\zeta(\xi)\,d\xi$$

then leads immediately to

$$\psi_r(a) = -y^{(1)}_r(a)\int_a^b y^{(2)}(\xi)\,\zeta(\xi)\,d\xi,$$

and

$$\psi_r(b) = -y^{(2)}_r(b)\int_a^b y^{(1)}(\xi)\,\zeta(\xi)\,d\xi.$$

Thus the solution to the Dirichlet problem for L also satisfies the Neumann boundary conditions provided the right hand side obeys the two compatibility conditions.

A straightforward approach to determining boundary conditions for $\zeta^{(1)}$ would be to use a closed quadrature formula for which the values $\zeta^{(1)}_0$ and $\zeta^{(1)}_N$ appear explicitly in order to solve for these two values from the two integral constraints. In our case this is unnatural, however, since the discretized version (10a) of the operator involves only the *interior* values of $\zeta^{(1)}_j$, so only these values should appear in the discrete version of the constraint conditions. Instead, we use the vorticity equation (10b) and the two conditions

$$0 = \frac{d}{dt}\int_a^b y^{(p)}(r)\,\zeta^{(1)}(r,t)\,dr$$

to determine boundary values such that the integrals are preserved in time; then if the constraints are satisfied by the initial data, we have a consistent formulation.

For a numerical procedure, we determine two compatibility conditions on the interior values of $\zeta_j^{(1)}$,

$$\sum_{j=1}^{N-1} y_j^{(p)} \zeta_j^{(1)} = 0 \tag{11}$$

as described below. We then require that

$$0 = \sum_{j=1}^{N-1} y_j^{(p)} \frac{\partial \zeta_j^{(1)}}{\partial t} = A_0^{(p)} \zeta_0^{(1)} + A_N^{(p)} \zeta_N^{(1)} + \sum_{k=1}^{N-1} A_k^{(p)} \zeta_k^{(1)} - \sum_{j=1}^{N-1} \left[a^2 + \frac{1}{r_j^2} \right] \zeta_j^{(1)} + 2ia \sum_{j=1}^{N-1} \frac{v_j^{(0)}}{r_j} v_j^{(1)}, \tag{12}$$

where

$$A_k^{(p)} = \sum_{j=1}^{N-1} y_j^{(p)} \left(4D_{jk}^{(2)} + \frac{2}{r_j} D_{jk} \right)$$

for $p = 1, 2$; this relation can be solved to give the desired relations for $\zeta_0^{(1)}$ and $\zeta_N^{(1)}$ in terms of interior values of the other data.

The two compatibility conditions are found as follows. The $N-1$ linear equations (10a) plus discrete versions of the four homogeneous boundary conditions $\psi^{(1)} = \psi_r^{(1)} = 0$ are written as matrix system of $N+3$ equations in the $N+1$ unknowns $\psi_j^{(1)}$, for $j = 0, \ldots, N$. Two linearly independent vectors of length $N+3$ that lie in the null space of the transpose of this matrix were determined numerically by means of the QR decomposition [43] of the transpose matrix; routine SQRDC in the LINPACK library was used [41]. Setting the inner product of the null vectors with the right hand side of the augmented $N+3$ by $N+1$ system to zero results in the two equations (11). The weights $y_k^{(p)}$ can be determined once and for all in a preprocessing step; then at each time step, the inner products with the weights are formed and the two equations (12) are solved for the boundary data. The two constraints (11) can be monitored as a function of time as an error check. To obtain consistent initial data, we choose $\psi^{(1)}$ at time $t = 0$ to be a low order polynomial in r that satisfies all four homogeneous boundary conditions, and determine the initial data for $\zeta_j^{(1)}$ by Eqn. (10a).

REFERENCES

[1] D. T. J. Hurle and E. Jakeman, *Introduction to the techniques of crystal growth*, PCH PhysicoChem. Hydrodyn. 2 (1981), pp. 237–244.

[2] M. E. Glicksman, E. Winsa, R. C. Hahn, T. A. Lograsso, S. H. Tirmizi, and M. E. Selleck, *Isothermal dendritic growth – a proposed microgravity experiment*, Metall. Trans. 19A (1988), pp. 1945–1953.

[3] R. A. Brown, *Theory of transport processes in single crystal growth from the melt*, AIChE J. 34 (1988), pp. 881-911.

[4] S. R. Coriell, G. B. McFadden and R. F. Sekerka, *Cellular growth during directional solidification*, Ann. Rev. Mater. Sci. 15 (1985), pp. 119–145.

[5] M. E. Glicksman, S. R. Coriell and G. B. McFadden, *Interaction of flows with the crystal-melt interface*, Annu. Rev. Fluid Mech. 18 (1986), pp. 307-335.

[6] W. W. Mullins and R. F. Sekerka, *Stability of a planar interface during solidification of a dilute binary alloy*, J. Appl. Phys. 35 (1964), pp. 444-451.

[7] S. R. Coriell, M. R. Cordes, W. J. Boettinger, and R. F. Sekerka, *Convective and interfacial instabilities during unidirectional solidification of a binary alloy*, J. Crystal Growth 49 (1980), pp. 13-28.

[8] S. R. Coriell and R. F. Sekerka, *Effect of convective flow on morphological stability*, PCH PhysicoChem. Hydrodyn. 2 (1981), pp. 281–293.

[9] S. H. DAVIS, *Hydrodynamic interactions in directional solidification*, J. Fluid. Mech. 212 (1990) pp. 241–262.
[10] R. T. DELVES, *Theory of Interface Stability*, in *Crystal Growth*, B. R. Pamplin, ed., Pergamon, Oxford, 1974, pp. 40-103.
[11] S. R. CORIELL, G. B. MCFADDEN, R. F. BOISVERT, AND R. F. SEKERKA, *Effect of a forced Couette flow on coupled convective and morphological instabilities during unidirectional solidification*, J. Crystal Growth 69 (1984), pp. 15-22.
[12] S. R. CORIELL AND G. B. MCFADDEN, *Buoyancy effects on morphological instability during directional solidification*, J. Crystal Growth 94 (1989), pp. 513-521.
[13] K. BRATTKUS, AND S. H. DAVIS, *Flow induced morphological instability: stagnation point flows*, J. Crystal Growth 89 (1988), pp. 423-427.
[14] G. B. MCFADDEN, S. R. CORIELL, AND J. I. D. ALEXANDER, *Hydrodynamic and free boundary instabilities during crystal growth: the effect of a plane stagnation flow*, Comm. Pure and Appl. Math 41 (1988), pp. 683-706.
[15] K. BRATTKUS AND S. H. DAVIS, *Flow induced morphological instability: the rotating disk*, J. Crystal Growth 87 (1988), pp. 385-396.
[16] S. A. FORTH AND A. A. WHEELER, *Hydrodynamic and morphological stability of the unidirectional solidification of a freezing binary alloy: a simple model*, J. Fluid Mech. 202 (1989), pp. 339-366.
[17] S. H. DAVIS, U. MÜLLER, AND C. DIETSCHE, *Pattern selection in single-component systems coupling Bénard convection and solidification*, J. Fluid. Mech. 144 (1984) pp. 133–151.
[18] B. CAROLI, C. CAROLI, C. MISBAH, AND B. ROULET, *Solutal convection and morphological instability in directional solidification of binary alloys*, J. Phys. (Paris) 46 (1985), pp. 401-413.
[19] G. W. YOUNG AND S. H. DAVIS, *Directional solidification with buoyancy in systems with small segregation coefficient*, Phys. Rev. B34 (1986), pp. 3388-3396.
[20] G. B. MCFADDEN, S. R. CORIELL, R. F. BOISVERT, M. E. GLICKSMAN, AND Q. T. FANG, *Morphological stability in the presence of fluid flow in the melt*, Metall. Trans. 15A (1984), pp. 2117-2124.
[21] Q. T. FANG, M. E. GLICKSMAN, S. R. CORIELL, G. B. MCFADDEN, AND R. F. BOISVERT, *Convective influence on the stability of a crystal-melt interface*, J. Fluid Mech. 151 (1985) pp. 121-140.
[22] G. B. MCFADDEN, S. R. CORIELL, M. E. GLICKSMAN, AND M. E. SELLECK, *Instability of a Taylor-Couette flow interacting with a crystal-melt interface*, PCH PhysicoChem. Hydrodyn. 11 (1989), pp. 387-409.
[23] G. B. MCFADDEN, S. R. CORIELL, B. T. MURRAY, M. E. GLICKSMAN, AND M. E. SELLECK, *Effect of a crystal-melt interface on Taylor-vortex flow*, Phys. Fluids A 2 (1990), pp. 700-705.
[24] G. I. TAYLOR, *Stability of a viscous liquid contained between two rotating cylinders*, Phil. Trans. Roy. Soc. A 223 (1923), 289-343.
[25] S. CHANDRASEKHAR, *Hydrodynamic and Hydromagnetic Stability*, (Clarendon Press, Oxford, 1961).
[26] P. G. DRAZIN AND W. H. REID, *Hydrodynamic Stability*, (Cambridge University Press, New York, 1981).
[27] E. A. CODDINGTON AND N. LEVINSON, *Theory of Ordinary Differential Equations*, (McGraw Hill, New York, 1955).
[28] S. CARMI AND J. I. TUSTANIWSKYJ, *Stability of modulated finite-gap cylindrical Couette flow: linear theory*, Journal of Fluid Mech. 108 (1981) 19–42.
[29] P. HALL, *The stability of unsteady cylinder flows*, J. Fluid Mech. 67 (1975) 29-63.
[30] G. SEMINARA AND P. HALL, *Centrifugal instability of a Stokes layer: linear theory*, Proc. R. Soc. Lond. A 350 (1976), 299–316.
[31] B. T. MURRAY, G. B. MCFADDEN, AND S. R. CORIELL, *Stabilization of Taylor-Couette flow due to time-periodic outer cylinder oscillation*, Phys. Fluids A, 2 (1990), 2147–2156.
[32] B. T. MURRAY, S. R. CORIELL, AND G. B. MCFADDEN, *The effect of gravity modulation on solutal convection during directional solidification*, J. Crystal Growth, in press.
[33] S. H. DAVIS, *The stability of time-periodic flows*, Annu. Rev. Fluid Mech. 8 (1976), 57–74.
[34] M. R. SCOTT AND H. A. WATTS, *Computational solution of linear two-point boundary value problems via orthonormalization*, SIAM J. Numer. Anal. 14 (1977), 40-70.

[35] SLATEC Common Math Library, National Energy Software Center, Argonne National Laboratory, Argonne, IL.

[36] H. B. Keller, *Numerical Solution of Two Point Boundary Value Problems*, Regional Conference Series in Applied Mathematics 24, SIAM, Philadelphia, 1976.

[37] The routine SNSQ is part of the SLATEC Common Math Library [35], and was written by K. L. Hiebert (1980); it is based on Powell [38].

[38] M. J. D. Powell, in *Numerical Methods for Nonlinear Algebraic Equations*, P. Rabinowitz, ed. (Gordon and Breach, New York, 1970), pp. 87–161.

[39] C. Canuto, M. Y. Hussaini, A, Quarteroni, and T. A. Zang, *Spectral Methods in Fluid Mechanics*, (Springer, New York, 1988).

[40] D. Gottlieb, M. Y. Hussaini and S. A. Orszag, in *Spectral Methods for Partial Differential Equations*, R. G. Voigt, D. Gottlieb, and M. Y. Hussaini, eds., (SIAM, Philadelphia, 1984) pp. 1–54.

[41] J. J. Dongarra, J. R. Bunch, C. B. Moler, G. W. Stewart, *LINPACK User's Guide*, (SIAM, Philadelphia, 1979).

[42] D. Kahaner, C. Moler, and S. Nash, *Numerical Methods and Software*, (Prentice Hall, New Jersey, 1989) p. 295.

[43] G. H. Golub and C. F. Van Loan, *Matrix Computations*, 2nd Ed., (Johns Hopkins University Press, Baltimore, 1989).

[44] G. B. McFadden, B. T. Murray, S. R. Coriell, M. E. Glicksman, and M. E. Selleck, *Effect of a crystal-melt interface on Taylor-Couette flow with buoyancy*, in *Proceedings of the 5th International Colloquium on Free Boundary Problems: Theory and Applications*, J. Chadam, ed., held in Montreal, Canada, June 13-22, 1990.

[45] G. B. McFadden, S. R. Coriell, R. F. Boisvert, and M. E. Glicksman, *Asymmetric instabilities in buoyancy-driven flow in a tall vertical annulus*, Phys. Fluids 27 (1984), pp. 1359–1361.

[46] L. Quartapelle, *Vorticity conditioning in the computation of two-dimensional viscous flows*, J. Comp. Phys. 40 (1981) pp. 453–477.

[47] C. R. Anderson, *Vorticity boundary conditions and boundary vorticity generation for two-dimensional viscous incompressible flows*, J. Comp. Phys. 80 (1989) 72-97.

A ONE DIMENSIONAL STOCHASTIC MODEL OF COARSENING

W.W. MULLINS*

Abstract. A one dimensional model of the coarsening of intervals on a line is considered in which the boundary points between adjacent intervals execute independent random walk with a common diffusion coefficient $D/2$; when two boundary points meet, they coalesce into a single point that continues to execute random walk. We calculate the following quantities in the asymptotic limit of long times: 1) The average interval length (i.e., $\langle l \rangle = \sqrt{\pi D t}$), 2) the time-independent probability density for the reduced length $\sigma = 1/\langle l \rangle$, and 3) the expected value of dl/dt for a given l, which is positive for $l > l_c = \sqrt{2/\pi}\ \langle l \rangle$ and negative for $l < l_c$. The model is similar to one proposed by Louat for grain growth. Although it is not a good representation of the details of most physical processes of coarsening, it is of theoretical interest since it is one of the few cases for which analytic results can be obtained.

1. Introduction. Coarsening is a process whereby a nearly fixed total volume of material is redistributed over a decreasing number of domains, as smaller domains feed material to larger domains and ultimately disappear. Hence the average volume of the surviving domains increases with time and the configuration of the system is said to coarsen. The domains may occupy the entire space of the system (grains in a polycrystal) or may be embedded in a matrix (precipitate coarsening). The process is driven by the reduction of the free energy (area in the simplest cases) of the interfaces between domains or between domains and the matrix.

In this paper, we consider a one dimensional model of coarsening in which the domains are intervals on a line (with no gaps) and the domain boundaries are points between adjacent intervals. The model is a stochastic one in which the domain boundaries are assumed to execute independent random walk, with a common diffusion coefficient D'. When two boundary points meet, they coalesce and the resulting single point continues to execute random walk with the same D'. This causes the configuration of intervals to coarsen in the manner depicted in Fig. 1, in which the vertical axis is time.

The model is similar to a three dimensional model of grain growth by random boundary fluctuations proposed earlier by Louat [1]. Although his model is not an appropriate one for ordinary grain growth, since a given boundary moves in response to macroscopic thermodynamic driving forces with negligible random fluctuations, it is one of the few models of coarsening for which analytic results can be obtained. Other models of coarsening that are amenable to analysis are those that can be treated by the classical theory of Lifshitz and Slyozov [2] and of Wagner [3].

*Carnegie Mellon University, Department of Metallurgical Engineering and Materials Science, 8309 Wean Hall, Pittsburgh, PA 15213

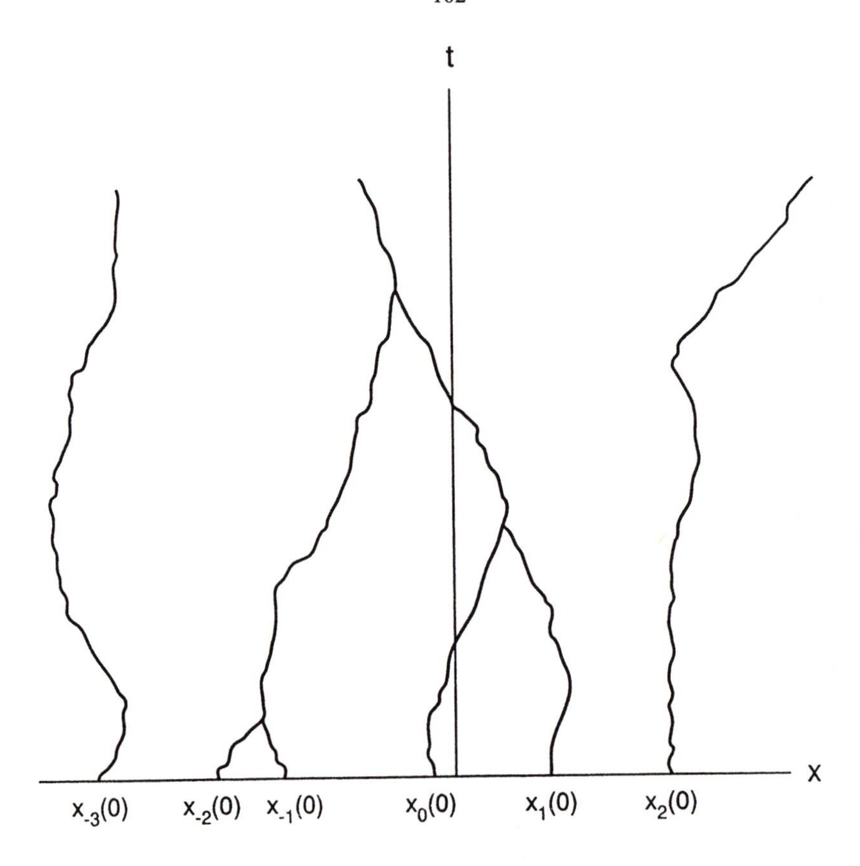

Fig. 1 Stochastic coarsening by random walk of domain boundaries

The results we obtain for the stochastic model show self-similar scaling properties in the asymptotic limit of long times. A system undergoing coarsening can be said to be in a regime of statistical self-similarity if any two configurations developed by the system in this regime are statistically indistinguishable when brought to the same linear scale by uniform magnification; thus any statistical distribution function expressed in terms of reduced lengths (length divided by average length) will be independent of time. The model we consider is one of the few models of coarsening for which one can prove that the asymptotic distribution of reduced interval lengths at long times is independent of time. It therefore provides a valuable example of the self-similar behavior observed in many coarsening systems [4,5].

2. Analysis of the stochastic model. We consider an infinite line with boundary points x_i that execute independent random walk with a diffusion coefficient D', as illustrated in Fig. 1. When two boundary points coalesce, their x coordinates remain equal thereafter. It follows that the ith interval $l_i = x_{i+1} - x_i$, if non-zero, also executes random fluctuations with a diffusion coefficient of $D = 2D'$, since for independent x_i, $\langle l_i^2 \rangle = 2\langle x_i^2 \rangle$. We define $Q(+ \cap l, t)dl$ to be the probability that an interval is non-zero (+) and lies in the range l to $l + dl$ at time t. In view of the foregoing assumptions, we see that Q satisfies the diffusion equation

$$\partial Q/\partial t = D\partial^2 Q/\partial l^2 \ , \tag{1}$$

with an absorbing boundary condition $Q(+ \cap 0, t) = 0$, since the interval becomes and remains zero at $l = 0$.

The solution of Eq.(1) with the absorbing boundary condition and the initial condition $l = l_0$ at $t = 0$ is well known and is given by the method of images as

$$Q(+ \cap l, t) = \frac{1}{(4\pi D t)^{1/2}} \left\{ \exp\left[-\frac{(l - l_0)^2}{4Dt} \right] - \exp\left[-\frac{(l + l_0)^2}{4Dt} \right] \right\} . \tag{2}$$

To find the probability $Q_0(+, t)$ that the interval has survived (i.e. is non-zero) up to time t, we integrate Eq.(2) over all l to obtain, as in the gambler's ruin problem,

$$Q_0(+, t) = \int_0^\infty Q(+ \cap l, t) dl = \operatorname{erf}\left[\frac{l_0}{2\sqrt{Dt}} \right] , \tag{3}$$

where $\operatorname{erf}[u] = (2/\sqrt{\pi}) \int_0^u \exp[-y^2] dy$. Eq.(3) shows that the survival probability of any interval initially present becomes arbitrarily small at sufficiently long times. The conditional probability density $P(l, t|+) = Q(+\cap l, t)/Q_0(+, t)$ for l of surviving intervals at t is given from Eqs.(2) and (3) by

$$P(l, t|+) = \frac{1}{(4\pi D t)^{1/2} \operatorname{erf}\{l_0/2\sqrt{Dt}\}} \left\{ \exp\left[-\frac{(l - l_0)^2}{4Dt} \right] - \exp\left[-\frac{(1 + l_0)^2}{4Dt} \right] \right\} . \tag{4}$$

If $4Dt \gg l_0^2$, then Eq.(4) becomes approximately

$$P(l, t|+) = \frac{l_2}{2Dt} \exp\left[-\frac{l^2}{4Dt} \right] , \tag{5}$$

which is normalized. Equation (5) shows that P is independent of l_0.

The average value of l at long times is given from Eq.(5) by

$$\langle l \rangle = \int_0^\infty l P(l, t|+) dl = \sqrt{\pi D t} \tag{6}$$

If we define a reduced length $\sigma = l/\langle l \rangle$, then the probability density $\mathcal{P}$ for σ is given from Eqs.(5) and (6) as (omitting the +)

$$\mathcal{P}(\sigma) = \frac{\pi}{2} \, \sigma \exp\left[-\frac{\pi}{4} \, \sigma^2 \right] . \tag{7}$$

The asymptotic density for the reduced length σ given by Eq.(7) is independent of time and is an example of what we have called statistical self-similarity. Clearly this density is developed asymptotically and never actually achieved in a finite time.

To calculate $\langle \dot{l}|l\rangle$, the expected value of $\dot{l} = dl/dt$ for a given l, we first define the number $n_L(l,t|+) = N_L P(l,t|+)$ of surviving intervals of length l per unit length, where $N_L = 1/\langle l\rangle$ is the number of intervals per unit length. It is easily verified that n_L satisfies the diffusion equation (i.e. Eq.(1)). Hence the flux of intervals J along an abstract l axis may be written $J = n_L\langle \dot{l}|l\rangle = -D\partial n_L/\partial l$. Solving for $\langle \dot{l}|l\rangle$, using the definition of n_L and cancelling N_L, we have

$$\langle \dot{l}|l\rangle = -D\frac{\partial lnP}{\partial l} \ . \tag{8}$$

Substituting the asymptotic expression for P from Eq.(5) into Eq.(8) and using the definition σ, we find

$$\langle l\rangle\langle \dot{l}|l\rangle = D\left(\frac{\pi\sigma}{2} - \frac{1}{\sigma}\right) \tag{9}$$

Equation (9) shows that if $\sigma > \sigma_c$, where the critical value σ_c is given by $\sigma_c = \sqrt{2/\pi}$, then $\langle \dot{l}|l\rangle$ is positive, whereas if $\sigma < \sigma_c$, then $\langle \dot{l}|l\rangle$ is negative. In this sense, large intervals tend to grow and small ones tend to shrink. As already noted, however, any given interval eventually shrinks and disappears. This happens essentially because the average interval length $\langle l\rangle$ continues to grow so that eventually the critical length $(l_c = \sigma_c\langle l\rangle)$ exceeds the length of any given interval.

3. Conclusions. We have obtained the following results for the one dimensional stochastic model of coarsening in which domain boundaries execute independent random walk:

1. The survival probability of any given domain becomes arbitrarily small for sufficiently long times (Eq.(3)).

2. The asymptotic probability density for the reduced domain length $\sigma = l/\langle l\rangle$ is independent of time (Eq.(7)).

3. The asymptotic average domain length is given by $\langle l\rangle = \sqrt{\pi Dt}$ (Eq.(6)).

4. In the asymptotic range, intervals larger than $l_c = \sqrt{2/\pi}\ \langle l\rangle$ grow on the average and intervals smaller than l_c shrink on the average (Eq.9).

Conclusions corresponding to 2 and 3 were obtained by Louat [1] for a less precisely specified model. The model is not a good representation of most physical processes of coarsening for which the motion of individual domain boundaries is not stochastic but rather is determined by physical parameters (e.g. boundary curvature, energy etc.). The model is, however, of theoretical interest because of the analytic results that can be obtained; in particular, it illustrates the phenomenon of statistical self-similarity that is observed in many systems.

Equation Eq.(9) shows that $\langle \dot{l}|l \rangle$ is inversely proportional to the scale of the configuration of intervals in the following sense: if the entire line is uniformly magnified by the scale factor λ so that new lengths are given by $l_\lambda = \lambda l$, then $\langle \dot{l}|l \rangle_\lambda = \frac{1}{\lambda} \langle \dot{l}|l \rangle$. This result together with conclusion 2 above, implies conclusion 3 according to a previous analysis [6].

It is clear that separate intervals are not statistically independent. For example, if the motion of a boundary between two adjacent intervals adds length to one interval, it must subtract length from the other one. The investigation of the correlations between intervals and their scaling behavior would be of considerable interest. If statistical self-similarity holds in the strong form, then all correlations when expressed in terms of reduced lengths will be independent of time.

Acknowledgement. This research was carried out for a conference supported by the Institute for Mathematics and its Applications with funds provided by the National Science Foundation.

REFERENCES

[1] N.P. Louat, Acta Metall. 22 (1974), 721.
[2] I.M. Lifshitz and V.V. Slyozov, J. Phys. Chem. Solids 19 (1961), 35.
[3] C. Wagner, Z. Electrochem. 65 (1961), 581.
[4] W.W. Mullins and J. Viñals, Acta Metall. 37 (1989), 991.
[5] P.W. Voorhees, J. Stat. Phys. 38 (1985), 231.
[6] W.W. Mullins, J. Appl. Phys. 59 (1986), 1341.

ALGORITHMS FOR COMPUTING CRYSTAL GROWTH AND DENDRITIC SOLIDIFICATION*

JAMES A. SETHIAN† AND JOHN STRAIN‡

Abstract. We report on a numerical method for computing the motion of complex solid/liquid boundaries in crystal growth. The model we solve includes physical effects such as crystalline anisotropy, surface tension, molecular kinetics and undercooling. The method is based on a single history-dependent boundary integral equation on the solid/liquid boundary, which is solved by means of a fast algorithm coupled to a level set approach for tracking the evolving boundary. Numerical experiments show the evolution of complex crystalline shapes, development of large spikes and corners, dendrite formation and side-branching, and pieces of solid merging and breaking off freely.

In this paper, we report on a numerical method for crystal growth and solidification. The physical motivation for this problem is as follows. Begin with a container of the liquid phase of the material under study, water for example. Suppose the box is smoothly and uniformly cooled to a temperature below its freezing point, so carefully that the liquid does not freeze. The system is now in a "metastable" state, where a small disturbance — such as dropping a tiny seed of the solid phase into the liquid — will initiate a rapid and unstable process known as dendritic solidification. The solid phase will grow from the seed by sending out branching fingers into the distant cooler liquid nearer the undercooled wall. This growth process is unstable in the sense that small perturbations of the initial data can produce large changes in the time-dependent solid/liquid boundary.

Mathematically, this phenomenon can be modeled by a moving boundary problem. The temperature field satisfies a heat equation in each phase, coupled through two boundary conditions on the unknown moving solid/liquid boundary, as well as initial and boundary conditions. The moving boundary conditions explicitly involve geometric properties of the boundary itself, such as the local curvature and the normal direction, as well as the temperature field. For further details, see [Cahn & Hilliard(2), Gurtin(6), Mullins & Serkerka(14), Mullins & Serkerka(15)].

A variety of techniques are possible to numerically approximate the equations of motion. One approach is to solve the heat equation numerically in each phase and try to move the boundary so that the two boundary conditions are satisfied, see [Chorin(3), Smith(22), Kelly & Ungar(8), Meyer(12), Sullivan et.al.(28)]. However, it is difficult to impose the boundary conditions accurately on a time-dependent and complicated boundary. Thus these calculations have been used mainly to study small perturbations of smooth crystal shapes. Another approach is to recast the

*The first author was supported in part by the Applied Mathematics Subprogram of the Office of Energy Research under contract DE-AC03-76SF00098. This author also acknowledges the support of the National Science Foundation under contract DMS89-19074 The second author was supported by DARPA/AFOSR Contract No. F-49620-87-C-0065 and a National Science Foundation Mathematical Sciences Postdoctoral Research Fellowship.

†Department of Mathematics, University of California, Berkeley, California 94720

‡Department of Mathematics, Princeton University, Princeton, New Jersey 08544

equations of motion as a single integral equation on the moving boundary and solve the integral equation numerically, as is done in [Meiron(11), Strain(24), Karma(7), Kessler & Levine(9), Langer(10)]. This approach can yield more accurate results for smooth boundaries, as well as agreement with linear stability theory. On the other hand, a parametrization of the boundary must be computed at each time step. The curvature and normal vector are then derivatives of the parametrization, and these methods usually break down in the interesting cases where the boundary becomes complicated and loses smoothness, as discussed in [Sethian(17), Sethian(18)]. In particular, the calculations presented in [Strain(24)] indicate that corners and cusps may form and pieces of the boundary may intersect; in either case, the boundary cannot be parametrized by a single smooth function. Finally, we note that most known numerical methods for crystal growth are difficult to extend to problems in three space dimensions.

In this paper, we report on a numerical method for crystal growth problems which avoids these difficulties and computes complex crystal shapes. Our method follows growth from an initial seed crystal or crystals (of arbitrary shape, size and location), in a standard model which includes the crystalline anisotropy, surface tension and molecular kinetics of the material, the undercooling imposed on the container walls, and the initial state of the crystal/temperature system. Our numerical calculations exhibit complicated shapes with spikes and corners, topological changes in the solid-liquid boundary, dendrite formation and sidebranching. The method relies on two main ideas. First, we represent the solid/liquid boundary as the zero set of a function ϕ defined on the whole container. The boundary is then moved by solving a nonlinear pseudo-differential equation suggestive of a Hamilton–Jacobi equation on the whole container. This level set "Hamilton–Jacobi" formulation of moving interfaces was introduced in [Osher & Sethian(16)], and allows us to compute geometric properties (see [Sethian(18)]) of highly complicated boundaries without relying on a parametrization. Hence, the moving boundary can develop corners and cusps and undergo topological changes quite naturally. Second, we reformulate the equations of motion as a boundary integral equation for the normal velocity as is also done in [Langer(10), Meiron(11), Strain(23), Kessler & Levine(9), Brush & Sekerka(1)]. We then extend the normal velocity smoothly to the whole container, as required by the level set Osher–Sethian algorithm. These two ideas combine to yield a fixed-domain formulation of the equations of motion, which may be of analytical interest itself. Our numerical method is based on solving a regularization of this equation, with entropy-satisfying upwind differencing for the level set equation of motion (see [Sethian(18)]), and a fast algorithm for evaluating the normal velocity (see Strain(24)]). The method generalizes immediately to higher order accurate schemes and, more importantly, to three-dimensional problems. The complete technical details of this work were first presented in [Sethian & Strain(20)], and we refer the interested reader to that paper for a complete discussion. This paper summarizes the key points of that paper. Here, we focus on an overview of the technique and some of the computed results. The extension of this algorithm to three-space dimensions will appear in [Sethian & Strain(21)].

1. The model equations of motion. We wish to model the growth of a solid crystal from a seed in a undercooled liquid bath. We consider a model which includes the effects of undercooling, crystalline anisotropy, surface tension, molecular kinetics, and initial conditions. We now state the equations of motion. Consider a square container, $B = [0,1] \times [0,1]$, filled with the liquid and solid phases of some pure substance, The unknowns are the temperature $u(x,t)$ for $x \in B$, and the solid-liquid boundary $\Gamma(t)$. The temperature field u is taken to satisfy the heat equation in each phase (see Figure 1), together with an initial condition in B and boundary conditions on the container walls.

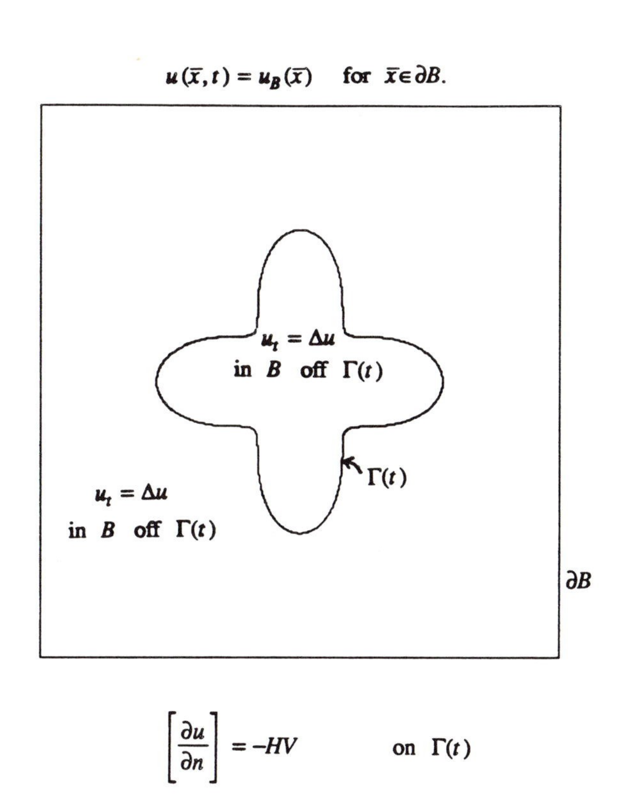

$$u(\bar{x},t) = -\epsilon_C(n)C - \epsilon_V(n)V \quad \text{for } \bar{x} \text{ on } \Gamma(t)$$

$$\epsilon_C(n) = \epsilon_C(1 - A\cos(k_A\theta + \theta_0))$$

$$\epsilon_V(n) = \epsilon_V(1 - A\cos(k_A\theta + \theta_0))$$

Figure 1: Equations of Motion

Thus

$$
\begin{array}{llll}
(1.1) & u_t = \Delta u & \text{in} & B \quad \text{off} \quad \Gamma(t) \\
 & u(x,t) = u_0(x) & \text{in} & B \quad \text{at} \quad t = 0 \\
 & u(x,t) = u_B(x) & \text{for} & x \in \partial B.
\end{array}
$$

Since the position of the moving boundary $\Gamma(t)$ is unknown, two boundary conditions on $\Gamma(t)$ are required to determine u and $\Gamma(t)$. Let n be the outward normal to the boundary, pointing from solid to liquid. The first boundary condition is the classical Stefan condition:

$$
(1.2) \qquad \left[\frac{\partial u}{\partial n}\right] = -HV \qquad \text{on} \quad \Gamma(t)
$$

Here $[\partial u/\partial n]$ is the jump in the normal component of heat flux $\partial u/\partial n$ from solid to liquid across $\Gamma(t)$, V is the normal velocity of $\Gamma(t)$, taken positive if liquid is freezing, and H is the dimensionless latent heat of solidification, which is a constant. The signs of geometric quantities are chosen so that if $\partial u/\partial n < 0$ in the liquid phase and $\partial u/\partial n = 0$ in the solid phase, then $[\partial u/\partial n]$ is negative and $V > 0$, indicating that the solid phase is growing. Physically, this means that undercooling drives solid growth. The latent heat of solidification controls the balance between geometry and temperature effects. The second boundary condition on $\Gamma(t)$ is the classical Gibbs-Thomson relation, modified to include crystalline anisotropy and molecular kinetics as well as the surface tension:

$$
(1.3) \qquad u(x,t) = -\varepsilon_C(n)C - \varepsilon_V(n)V \qquad \text{for} \quad x \quad \text{on} \quad \Gamma(t)
$$

Here C is the curvature at x on $\Gamma(t)$, taken positive if the center of the osculating circle lies in the solid. The anisotropy functions are modeled by

$$
(1.4) \qquad \varepsilon_C(n) = \varepsilon_C(1 - A\cos(k_A\theta + \theta_0))
$$

$$
(1.5) \qquad \varepsilon_V(n) = \varepsilon_V(1 - A\cos(k_A\theta + \theta_0))
$$

where θ is the angle between n and the x-axis, and ε_C, ε_V, A, k_A and θ_0 are constants depending on the material and the experimental arrangement. For example, if $\varepsilon_C = 0$ ($\varepsilon_V = 0$), there are no surface tension (molecular kinetic) effects. For $A = 0$, the system is isotropic, while if $A > 0$, the solid is k_A-fold symmetric with a symmetry axis at angle θ_0 to the x-axis. Typically $A \leq 1$. The equations of motion under study are (1.1), (1.2) and (1.3), with anisotropy functions given by (1.4) and (1.5).

2. Outline of the numerical algorithm. In this section, we sketch the numerical method we use to solve the equations of motion. We present the algorithm in four steps. In the first two steps, we transform the equations of motion into a boundary integral formulation. In the last two, we describe the level set formulation for moving the boundary and the necessary extension of V. The details of each step may be found in the sections that follow. However, the fundamental philosophy behind the algorithm is most easily conveyed through a series of figures. In Figure 2a, we show a typical solid/liquid boundary.

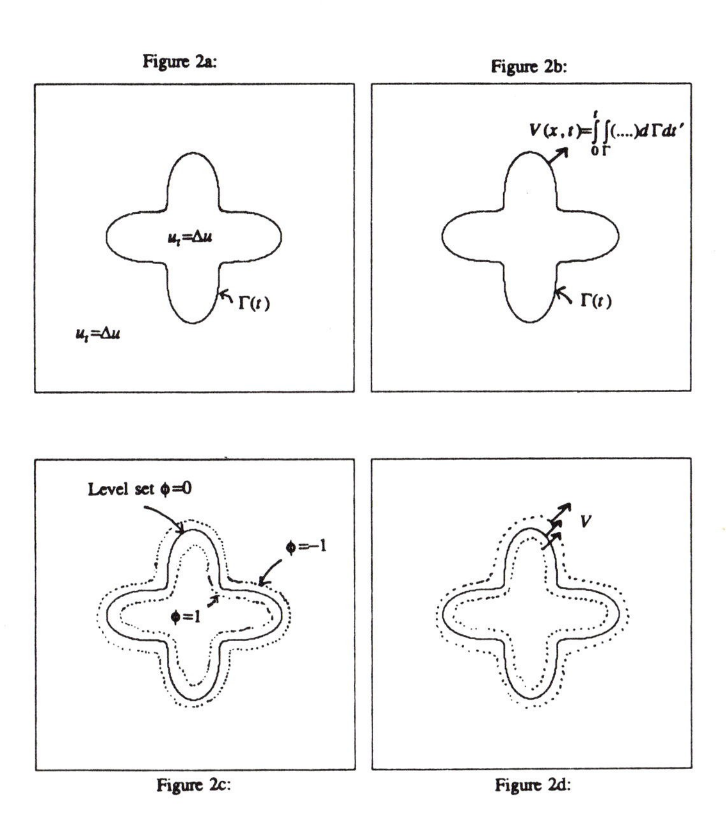

Figure 2a: P.D.E. for u in all space

Figure 2b: Integral equation for speed V of moving interface

Figure 2c: Level set function ϕ defined in all space

Level set $\phi = 0$ = solid/liquid boundary

Figure 2d: All level sets moved with velocity defined by integral equation

According to Eqns. (1.1-6), the temperature field u must satisfy the heat equation in each phase, as well as two boundary conditions on the phase boundary. In Steps 1 and 2, we transform the equations of motion into a single boundary integral equation on the moving boundary (see Figure 2b). Thus, we transform equations of motion which require computation of the temperature field u on the whole domain into an equation which involves only the moving boundary and its previous history.

In Step 3, we represent the boundary $\Gamma(t)$ as the level set $\{\phi = 0\}$ of a function ϕ defined on all of B (see Figure 2c). We construct a nonlinear pseudodifferential equation which evolves ϕ in such a way that the zero set $\{\phi = 0\}$, at each time t, is the moving boundary $\Gamma(t)$ (see Figure 2d). To do this, in Step 4 we construct a "speed function" F, which is equal to the normal velocity V on $\Gamma(t)$ and smoothly extends V to all of B. We then move all the level sets of ϕ with normal velocity F. The resulting level set equation for ϕ, which is reminiscent of a Hamilton-Jacobi equation, may be solved numerically by finite difference schemes borrowed from hyperbolic conservation laws. A major advantage of this formulation is that the equation for ϕ can be solved on a uniform mesh on the box B; the level sets are moved without constructing them explicitly. We now describe the algorithm in more detail.

Step 1: Subtraction of Initial and Boundary Conditions. Recall the equations of motion (1.1–6). The temperature field u and moving solid/liquid boundary $\Gamma(t)$ satisfy

$$\begin{aligned} &u_t = \Delta u \quad \text{in} \quad B - \Gamma(t) \\ u(x,t) = u_0(x) \quad \text{for} \quad t = 0 &\qquad u(x,t) = u_B(x) \quad \text{for} \quad x \in \partial B \end{aligned} \tag{2.1}$$

$$\left[\frac{\partial u}{\partial n}\right] = -HV \quad \text{on} \quad \Gamma(t) \tag{2.2}$$

$$u(x,t) = -\varepsilon_C(n)C - \varepsilon_V(n)V \quad \text{for} \quad x \quad \text{on} \quad \Gamma(t) \tag{2.3}$$

First, we subtract the temperature field due to the initial conditions u_0 and boundary conditions u_B. Let $U(x,t)$ be the solution to the heat equation

$$\begin{aligned} U_t &= \Delta U \quad \text{in} \quad B \\ U(x,0) &= u_0(x) \quad \text{at} \quad t = 0 \\ U(x,t) &= u_B(x,t) \quad \text{for} \quad x \in \partial B \quad \text{and} \quad t > 0. \end{aligned} \tag{2.4}$$

Let $W = u - U$. Then W satisfies

$$\begin{aligned} &W_t = \Delta W \quad \text{in} \quad B - \Gamma(t) \\ W(x,t) = 0 \quad \text{at} \quad t = 0 &\qquad W(x,t) = 0 \quad \text{for} \quad x \in \partial B \end{aligned} \tag{2.5}$$

$$\left[\frac{\partial W}{\partial n}\right] = -HV \quad \text{on} \quad \Gamma(t) \tag{2.6}$$

$$W(x,t) = -\varepsilon_C(n)C - \varepsilon_V(n)V - U(x,t) \quad \text{for} \quad x \quad \text{on} \quad \Gamma(t) \tag{2.7}$$

Eqns. (2.5–7) are equivalent to the original equations of motion. Given the solution U, at any time we can add W to U to produce the solution u to the original problem. The "free" temperature field $U(x,t)$, which is defined on a fixed domain, may be found analytically for simple initial and boundary conditions, or by numerical calculation in practical situations.

Step 2: Transformation to a Boundary Integral Equation. The next step is to transform Eqns. (2.5–7) into a integral equation on the boundary $\Gamma(t)$. Here, we follow the derivation given in [Strain(23)]. For further details, see [Strain(25), Strain(26)]. We use the kernel K of the heat equation to express the solution W to Eqns. (2.5–6) as a single layer heat potential. Given a function V on

$$\Gamma_T = \prod_0^T \Gamma(t) = \{(x,t) | x \in \Gamma(t),\ 0 \le t \le T\},$$

the single layer heat potential SV is defined for (x,t) in $B \times [0,T]$ by

$$SV(x,t) = \int_0^t \int_{\Gamma(t')} K(x, x', t - t') V(x', t') dx' dt'. \tag{2.8}$$

Here the x' integration is over the curves comprising $\Gamma(t')$, and the Green function K of the heat equation in the box $B = [0,1] \times [0,1]$ with Dirichlet boundary conditions on the box walls is given by

(2.9)

$$K(x, x', t) = \sum_{k_1=1}^{\infty} \sum_{k_2=1}^{\infty} e^{-(k_1^2+k_2^2)\pi^2 t} \sin(k_1 \pi x_1) \sin(k_2 \pi x_2) \sin(k_1 \pi x_1') \sin(k_2 \pi x_2')$$

(2.10)

$$= \frac{1}{4\pi t} \sum_{k_1=-\infty}^{\infty} \sum_{k_2=-\infty}^{\infty} \sum_{\sigma=\pm 1} \sum_{\sigma_2=\pm 1} \sigma_1 \sigma_2 e^{-[(x_1 - \sigma_1 x_1' - 2k_1)^2 + (x_2 - \sigma_2 x_2' - 2k_2)^2]/4t}$$

where $x = (x_1, x_2)$ and $x' = (x_1', x_2')$. The first expression for K can be calculated by Fourier series, the second by the method of images. The function $SV(x,t)$ defined by Eqn. (2.8) is a continuous function on $B \times [0,T]$, vanishing for $t = 0$ or x on ∂B, and satisfying the heat equation everywhere off Γ_T. Across $\Gamma(t)$, however, $SV(x,t)$ has a jump in its normal derivative equal to V. Thus, $W(x,t) = H \cdot SV(x,t)$ is the solution to Eqns. (2.5–6). All that remains is to satisfy the second boundary condition Eqn. (2.7). This is equivalent to the boundary integral equation

$$\varepsilon_C(n) C + \varepsilon_V(n) V + U + H \int_0^t \int_{\Gamma(t')} K(x, x', t - t') V(x', t') dx' dt' = 0 \tag{2.11}$$

for $x \in \Gamma(t)$. Eqn. (2.11) is an integral equation for the normal velocity of the moving boundary. We note that the velocity V of a point x on $\Gamma(t)$ depends not only on the position of $\Gamma(t)$ but also on its previous history. Thus, we have stored information about the temperature off the moving boundary in the previous history of the boundary.

Step 3: Level Set Formulation of Moving Boundaries. We follow the moving front by using the level set technology introduced in [Osher & Sethian(16)]. With this approach, a complex boundary can be advanced. Sharp corners and cusps are handled naturally, and changes of topology in the moving boundary require no additional effort. Furthermore, these methods work in any number of space dimensions. Recently, this technique has been applied to interface problems in mean curvature flow [Sethian(18)], singularity formation and minimal surface construction [Sethian & Chopp(19)], and compressible gas dynamics [Mulder & Osher & Sethian(13)]. In addition, theoretical analysis of mean curvature flow based on the level set model presented in [Osher & Sethian(16)] has recently been developed in [Evans & Spruck(4)]. We now describe the level set algorithm in the general case when a curve or union of curves $\Gamma(t)$ moves with speed V normal to itself. The essential idea is to construct a function $\phi(x,t)$ defined everywhere on B, such that the level set $\{\phi = 0\}$ is the set $\Gamma(t)$, that is,

$$\Gamma(t) = \{x : \phi(x,t) = 0\}. \tag{2.12}$$

We now derive a partial differential equation for ϕ, which holds on $B \times [0,T]$. Suppose we can construct a smooth function $F(x,t)$ defined on all of B such that

$$V(x,t) = F(x,t) \qquad \text{for} \quad x \in \Gamma(t) \tag{2.13}$$

Then $V = F$ on $\Gamma(t)$, and we call F a *smooth extension of V off* $\Gamma(t)$. We shall postpone until the next section the extension for the case of crystal growth. What is the equation of motion for ϕ? Obviously, this equation must reduce to normal propagation by speed V on the level set $\phi = 0$. Suppose we initialize $\phi(x,0)$ such that

$$\phi(x,0) = \pm \quad \text{distance from} \quad t \quad \text{to} \quad \Gamma(t) \tag{2.14}$$

where the plus (minus) sign is chosen if x is inside (outside) the initial boundary $\Gamma(t=0)$. Now consider the motion of some level set $\{\phi(x,t) = C\}$. Here, we follow the derivation given in [Mulder, Osher & Sethian(13)]. Let $x(t)$ be the trajectory of a particle located on this level set, so

$$\phi(x(t),t) = C \tag{2.15}$$

The particle speed $\frac{\partial x}{\partial t}$ in the direction n normal to $\Gamma(t)$ is given by the speed function F. Thus,

$$\frac{\partial x}{\partial t} \cdot n = F \tag{2.16}$$

where the normal vector n is given by $n = \frac{-\nabla\phi}{|\nabla\phi|}$. The minus sign is chosen because, given our initialization of ϕ, $\nabla\phi$ points inward. By the chain rule,

$$\phi_t + \frac{\partial x}{\partial t} \cdot \nabla\phi = 0 \tag{2.17}$$

and substitution yields

$$\begin{aligned} \phi_t - F|\nabla\phi| &= 0 \\ \phi(x, t=0) &= \text{ given} \end{aligned} \tag{2.18}$$

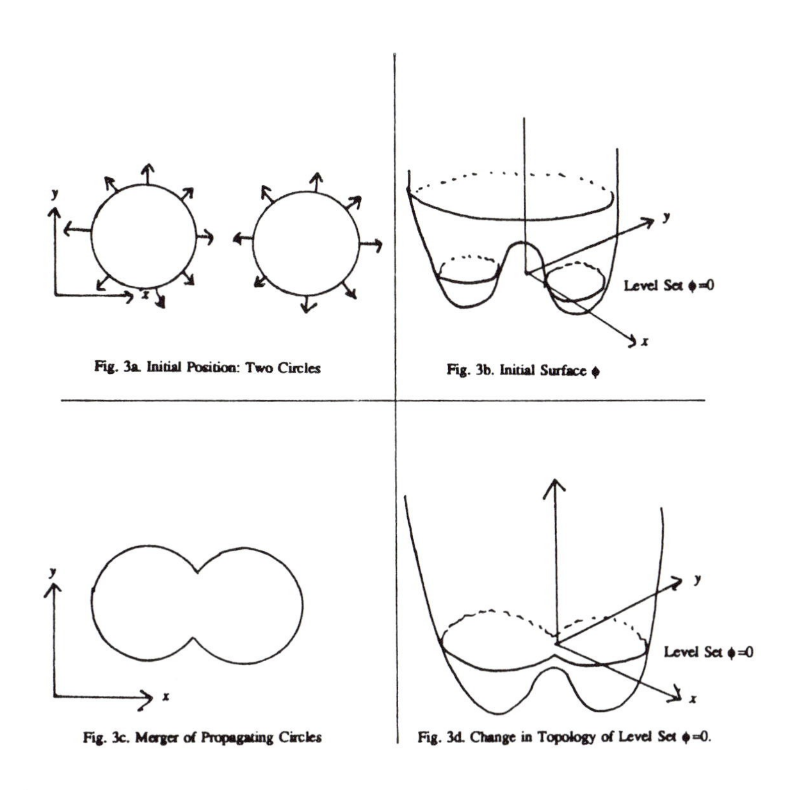

Figure 3: Hamilton–Jacobi Formulation

Equation (2.18) yields the motion of $\Gamma(t)$ with normal velocity V on the level set $\phi = 0$. We refer to Eqn. (2.18) as the level set "Hamilton–Jacobi" formulation. It is not strictly a Hamilton–Jacobi equation except for certain speed functions F, but the flavor of Hamilton–Jacobi equations is present. There are at least two advantages of this level set formulation compared to methods based on parametrizing $\Gamma(t)$. The first is that $\phi(x,t)$ always remains a function, even if the level surface $\phi = 0$ corresponding to the boundary $\Gamma(t)$ changes topology, breaks, merges or forms sharp corners. Parametrizations of the boundary become multiple-valued or singular in these cases. As an example, consider two circles in R^2 expanding outward with normal velocity $V = 1$ (Figure 3a). The initial function ϕ is a double-humped function which is continuous but not everywhere differentiable (Figure 3b). As ϕ evolves under Eqn. (4.2), the topology of the level set $\phi = 0$ changes. When the two circles expand, they meet and merge into a single closed curve with two corners

(Figure 3c). This is reflected in the change of topology of the level set $\phi = 0$ (Figure 3d). The initial value problem (2.18) can be solved numerically by finite difference schemes. Because ϕ can develop corners and sharp gradients, numerical techniques borrowed from hyperbolic conservation laws are used to produce upwind schemes for ϕ which track sharp corners accurately and employ the correct boundary conditions at the edge of the computational box.

Step 4: Construction of the Speed Function F. We now describe how to extend the velocity V to a globally defined speed function F. Such an extension is necessary in order to use the level set formulation. The most natural extension makes direct use of the integral equation (2.11), namely

$$\varepsilon_C(n)C + \varepsilon_V(n)V + U + H \int_0^t \int_{\Gamma(t')} K(x, x', t-t')V(x', t')dx'dt' = 0 \tag{2.19}$$

for $x \in \Gamma(t)$. Each term in Eqn. (2.19) can be evaluated anywhere in B, once V is known on $\Gamma(t')$ for $0 \leq t' \leq t$ and ϕ is known on B. Thus, given the set $\Gamma(t)$ plus all its previous positions and velocities for $0 \leq t' \leq t$, one could first solve an integral equation to find the velocity V for all points on $\Gamma(t)$ and then find $F(x,t)$ by solving the equation

$$\varepsilon_C(n)C + \varepsilon_V(n)F(x,t) + U(x,t) + H \int_0^t \int_{\Gamma(t')} K(x, x', t-t')V(x', t)dx'dt' = 0 \tag{2.20}$$

for F throughout B. Here, x is a point in B, while $C = C(x,t)$ is the curvature and n is the outward normal vector to the level set passing through x;

$$C = \nabla \cdot \left(\frac{\nabla \phi}{|\nabla \phi|}\right) = \nabla \cdot n \tag{2.21}$$
$$n = \frac{-\nabla \phi}{|\nabla \phi|}$$

and these expressions make sense everywhere in B. The speed function F given by Eqn. (2.20) is defined throughout B and equal to V on the solid-liquid boundary $\Gamma(t)$. For our purposes, a regularization of the speed function is easier to work with. We split the single layer heat potential into a history part $S_\delta V$ and a local part $S_L V$ as follows:

$$SV(x,t) = \int_0^{t-\delta} \int_{\Gamma(t')} K(x, x', t-t')V(x', t')dxdt' + \int_{t-\delta}^{t} \int_{\Gamma(t')} K(x, x', t-t')V(x', t')dx'dt'$$
$$\equiv S_\delta V + S_L V \tag{2.22}$$

Here δ is a small regularization parameter. Heuristically, we try to separate the local part, which is causing the jump in the normal derivative of the potential, from the history part which is smooth and independent of the current velocity. In [Sethian & Strain(20)], we show that the local part $S_L V$ can be approximated by

$$S_L V(x,t) = \sqrt{\frac{\delta}{\pi}}\; V(x,t) + O(\delta^{3/2}) \tag{2.23}$$

at points x on $\Gamma(t)$. The history part $S_\delta V$ depends only on values of V at times t' bounded away from the current time, since $t' \le t - \delta$ in $S_\delta V$, and is evaluated using a fast algorithm, see [Greengard & Strain(5), Strain(24)]. Since the leading term $\sqrt{\delta/\pi}\; V$ in the expression for the local part $S_L V$ varies smoothly in the direction normal to the curve if V does, this suggests the following regularized extension of V. Let F be defined everywhere in B by

$$\varepsilon_C(n)C + \varepsilon_V(n)F(x,t) + U(x,t) + H\sqrt{\frac{\delta}{\pi}}\; F(x,t) + H\; S_\delta V = 0 \tag{2.24}$$

We can then solve this equation for F to produce

$$F = \frac{-1}{\varepsilon_V(n) + H\sqrt{\delta/\pi}} \left[\varepsilon_C(n)C + U + H\; S_\delta V\right] \tag{2.25}$$

Thus, F is an explicit function of ϕ, the velocity at previous times $t' \le t - \delta$, and the free temperature field U. Note that F is defined even if $\varepsilon_V = 0$. Furthermore, F is equal to $V + O(\delta^{3/2})$ on $\Gamma(t)$, and does not have a boundary layer as we move off the curve, as long as δ is not too small. Finally, in the limit $\delta \to 0$, Eqn. (2.25) reduces to Eqn. (2.20) as it should.

3. Outline of the algorithm. In this section, we describe the general flow of the algorithm from one time step to the next. We have a pair of equations for the speed function F and the level set function ϕ defined throughout B, namely

$$\varepsilon_C(n)C + \varepsilon_V(n)F + U(x,t) + H\sqrt{\frac{\delta}{\pi}}\; F + H\; S_\delta V = 0 \tag{3.1}$$

$$\phi_t - F|\nabla\phi| = 0 \tag{3.2}$$

Here, the curvature C and the normal vector n are functionals of ϕ. U is computed by solving the heat equation on B. $S_\delta V$ is computed from the previous history of $\Gamma(t)$. To describe the algorithm, we imagine that at time step $n\Delta t$ we have (1) the level set function ϕ^n_{ij} defined at discrete grid points x_{ij}, (2) the free temperature field U^n_{ij} defined on the same discrete grid, and (3) the previous positions of the boundary $\Gamma(m\Delta t)$, $m = n-1, n-2, \ldots, n-d$ (where $d = \delta/\Delta t$), stored as points on each curve. Given this information, we proceed from one time step to the next as follows:

Step 1: At each grid point x_{ij}, compute the extended speed function F^n_{ij} from (3.1). This is done as follows:

(a)Expressions for the discrete curvature C_{ij} and normal vector n_{ij} may be computed from the discrete level set function ϕ_{ij}^n without explicit construction of the particular level set passing through the grid point x_{ij}, see [Sethian & Strain(20)].

(b) The history part $S_\delta V$ is updated by using the stored boundary at previous time levels.

Step 2: Calculate ϕ_{ij}^{n+1} from ϕ_{ij}^n and F_{ij}^n using the upwind, finite difference scheme described in [Sethian & Strain(20)].

Step 3: Calculate U_{ij}^{n+1} from U_{ij}^n by solving a finite difference approximation to the heat equation, [Sethian & Strain(20)].

Step 4: Find points on the new boundary $\Gamma((n+1)\Delta t)$ by constructing the level set $\phi = 0$ from ϕ_{ij}^{n+1}. Store the position x and velocity V of each point, found by interpolating F from the values on the grid. Note that these points on the boundary do not move. They serve only as quadrature points for updating the history integral $S_\delta V$.

Step 5: Replace n by $n+1$ and return to Step 1.

4. Numerical results. In this section, we present the results of our numerical calculations using our solidification algorithm. We begin with a series of calculations to check numerical consistency; that is, to verify that the computed solution converges as the numerical parameters are refined. (A further verification comparing the computed solution using the level set approach with an accurate boundary integral calculation in their common range of validity is in progress, see [27]). After demonstrating the robustness of the algorithm, we perform a study of the relative influence of the physical parameters, analyzing the effect of the surface tension coefficient ε_C, the kinetic coefficient ε_V, coefficient of anisotropy A, crystalline symmetry k_A, latent heat

1. Smooth growth. We begin with a calculation performed with surface tension coefficient $\varepsilon_C = .001$, kinetic coefficient $\varepsilon_V = .01$, no crystalline anisotropy (coefficient of anisotropy $A = 0$), and latent heat of solidification $H = 1$. The calculations were performed in a unit box, with a constant undercooling on the side walls of $u_B = -1$. The initial shape was a perturbed circle with average radius $R = .15$ and perturbation size $P = .08$ and $L = 4$ limbs. That is, the parametrized curve $(x(s), y(s))$, $0 \leq s \leq 1$ describing the initial position of the crystal is given by

$$(x(s), y(s)) = \left[(R + P\cos 2\pi Ls)\right](\cos 2\pi s, \sin 2\pi s) \tag{4.1}$$

with R= .15, $P = .08$, and $L = 4$. In Figure 4, we show a series of calculations performed to study the effect of refining the grid size and time step on the computed solution. We begin in Figure 4a with mesh size 48×48, and time step $\Delta t = .005$. The boundary grows smoothly outward from the initial limbs. Each limb is drawn towards its wall by the effect of the undercooling. The value of the kinetic coefficient ε_V is large enough to keep the evolving boundary smooth (in contrast to those calculations discussed below). In Figure 4b we repeat the above calculation with mesh size 64×64 and $\Delta t = .0025$. In Figure 4c, we take mesh size 96×96 and $\Delta t = .001$. The figures are unchanged, indicating the robustness of the algorithm. Although the shape remains smooth as it evolves, several effects can be seen. First, while the undercooling pulls the limbs towards the walls, the other walls act to thicken the limb, creating a highly smoothed version of side-branching. Second, the tips of the limbs, where the curvature is positive, move very fast compared to the indented pockets between limbs, where the curvature is negative.

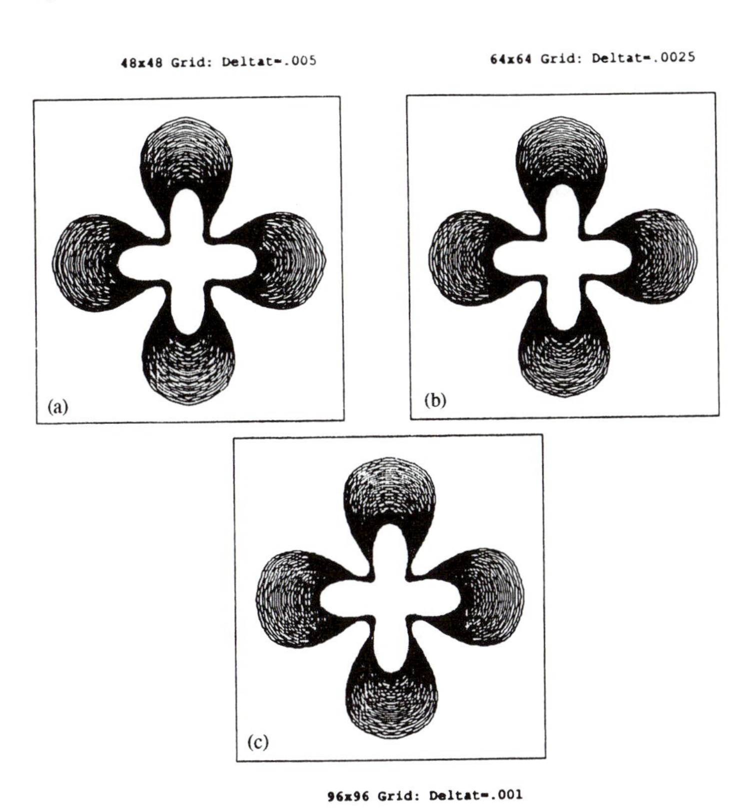

Figure 4: Smooth Crystal: Effect of Refining Both Grid Size and Time Step

Fig 4a: 48×48 Mesh, $\Delta t = 0.005, H = 1.0, A = 0.0, \varepsilon_C = 0.001, \varepsilon_V = 0.01, k_A = 0$

Fig 4b: 64×64 Mesh, $\Delta t = 0.0025, H = 1.0, A = 0.0, \varepsilon_C = 0.001, \varepsilon_V = 0.01, k_A = 0$

Fig 4c: 96×96 Mesh, $\Delta t = 0.001, H = 1.0, A = 0.0, \varepsilon_C = 0.001, \varepsilon_V = 0.01, k_A = 0$

2. Fingered growth

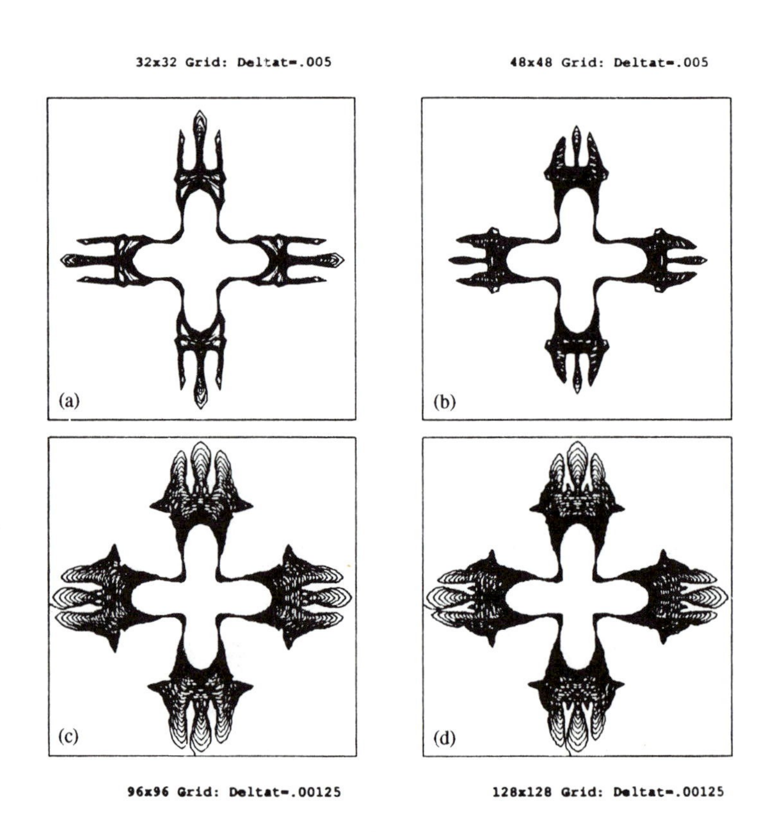

Figure 5: Fingered Crystal: Effect of Refining Both Grid Size and Time Step

Fig. 5a: 32×32 Mesh, $\Delta t = 0.005, H = 1.0, A = 0.0, \varepsilon_C = 0.001, \varepsilon_V = 0.001, k_A = 0$

Fig. 5b: 48×48 Mesh, $\Delta t = 0.005, H = 1.0, A = 0.0, \varepsilon_C = 0.001, \varepsilon_V = 0.001, k_A = 0$

Fig. 5c: 96×96 Mesh, $\Delta t = 0.005, H = 1.0, A = 0.0, \varepsilon_C = 0.001, \varepsilon_V = 0.001, k_A = 0$

Fig. 5d: 128×128 Mesh, $\Delta t = 0.005, H = 1.0, A = 0.0, \varepsilon_C = 0.001, \varepsilon_V = 0.001, k_A = 0$

Next, we perform a similar calculation, but change the kinetic coefficient to $\varepsilon_V = .001$. Once again, we take surface tension coefficient $\varepsilon_C = .001$, no crystalline anisotropy (coefficient of anisotropy $A = 0$), and latent heat of solidification $H = 1$. The constant undercooling on the side walls is -1, and the initial shape is again a perturbed circle with average radius .15, perturbation size .08, and 4 limbs, as in Eqn. (7.1). In Figure 5, we show the results using these values for the physical parameters. At the same time, we study the effects of refining the numerical parameters on the evolving calculations. In Figure 5a, we take a 32×32 mesh with $\Delta t = .005$. In Figure 5b, we take a 48×48 mesh with $\Delta t = .005$. In Figure 5c, we take a 96×96 mesh with $\Delta t = .00125$. In Figure 5d, we take a 128×128 mesh with

$\Delta t = .00125$. Starting from the smooth perturbed circle, the evolving boundary changes dramatically. First, each limb flattens out. Then, tip splitting occurs as spikes develop from each limb. Finally, side-branching begins as each multi-tipped arm is both pulled toward the closest wall and also drawn by the walls parallel to the limb. On the coarsest mesh (32×32), only the gross features of the fingering and tip splitting process are seen. As the numerical parameters are refined, the basic pattern emerges. It is clear that the resulting shapes are qualitatively the same, and there is little quantitative difference between Figures 5c and 5d. Even when computing the highly complex boundaries seen in these figures, the algorithm remains robust.

3. Latent heat of solidification

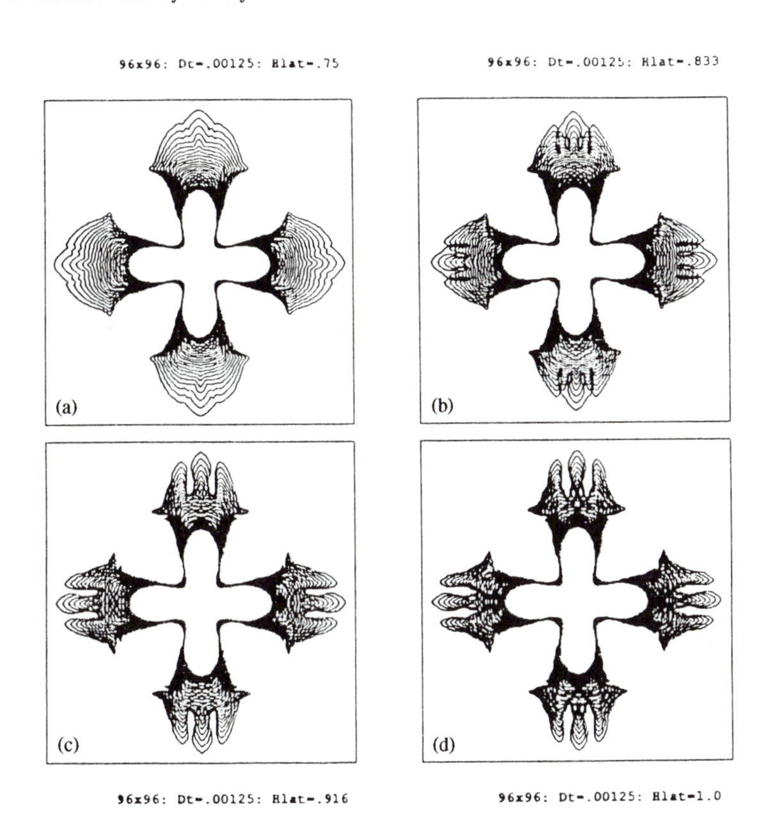

Figure 6: Small-Scale Refinement of Latent Heat of Solidification H

Fig. 6a: $H = 0.75, A = 0.0, \varepsilon_C = 0.001, \varepsilon_V = 0.001, k_A = 0, 96 \times 96, \Delta t = 0.00125$

Fig. 6b: $H = 0.833, A = 0.0, \varepsilon_C = 0.001, \varepsilon_V = 0.001, k_A = 0, 96 \times 96, \Delta t = 0.00125$

Fig. 6c: $H = 0.916, A = 0.0, \varepsilon_C = 0.001, \varepsilon_V = 0.001, k_A = 0, 96 \times 96, \Delta t = 0.00125$

Fig. 6d: $H = 1.0, A = 0.0, \varepsilon_C = 0.001, \varepsilon_V = 0.001, k_A = 0, 96 \times 96, \Delta t = 0.00125$

To understand the transition from smooth crystals to complex ones, we study the effect of changing the latent heat of solidification H. Recall that H controls the balance between the pure geometric effects and the solution of the history-dependent heat integral. We consider the same initial shape and physical parameters as in Figure 5 (again, $\varepsilon_V = .001$, $\varepsilon_C = .001$, $k_A = 0$, $A = 0$, with constant undercooling -1). In all calculations, we use a 96×96 mesh with time step $\Delta t = .00125$. In Figure 6 we perform a small-scale refinement of the latent heat of solidification for values $H = .75$ (Fig. 6a), $H = .833$ (Fig. 6b), $H = .916$ (Fig. 6c), and $H = 1.0$ (Fig. 6d). In this set of figures, the evolving boundaries are given at the same time. Instead, the final shape is plotted when $\Gamma(t)$ has reached to within .02 of the box walls (recall that the box has width 1.0). As the latent heat of solidification is increased, the growing limbs expand outwards less smoothly, and instead develop flat ends. As seen in Figure 6, these flat ends are unstable and serve as precursors to tip splitting. The mechanism operating here is presumably that increasing latent heat decreases the most unstable wavelength, as described by linear stability theory.

4. Changing initial conditions, anisotropy coefficient A and crystalline symmetry k_A. In Figures 7 and 8, we compute a collection of shapes by altering the initial seed, anisotropy coefficient, and degree of crystalline anisotropy. All calculations are performed using a 96×96 grid, with time step $\Delta t = 0.00125$, and $\varepsilon_V = 0.001$, $\varepsilon_C = 0.001$, $H = 1.0$, and constant undercooling -1. In Figure 7, we concentrate on changing the anisotropy coefficient and the crystalline anisotropy. In Fig. 7a, we take an initial shape with 4 limbs, four-fold crystalline anisotropy, and $A = 0.4$. In Fig. 7b, we take an initial shape with 4 limbs, six-fold crystalline anisotropy, and $A = 0.4$. In Fig. 7c, we take an initial shape with 4 limbs, four-fold crystalline anisotropy, and $A = 0.8$. In Fig. 7d, we take an initial shape with 4 limbs, six-fold crystalline anisotropy, and $A = 0.8$. The results depend dramatically on the choice of these values. In Figure 8, we concentrate on changing the anisotropy coefficient and the initial shape of the crystal. In Fig. 8a, we take an initial shape with 6 limbs, six-fold crystalline anisotropy, and $A = 0$. In Fig. 8b, we take an initial shape with 8 limbs, eight-fold crystalline anisotropy, and $A = 0$. In Fig. 8c, we take an initial shape with 6 limbs, six-fold crystalline anisotropy, and $A = 0.4$. In Fig. 8d, we take an initial shape with 8 limbs, eight-fold crystalline anisotropy, and $A = 0.4$. The resulting shapes display a variety of intricate behavior.

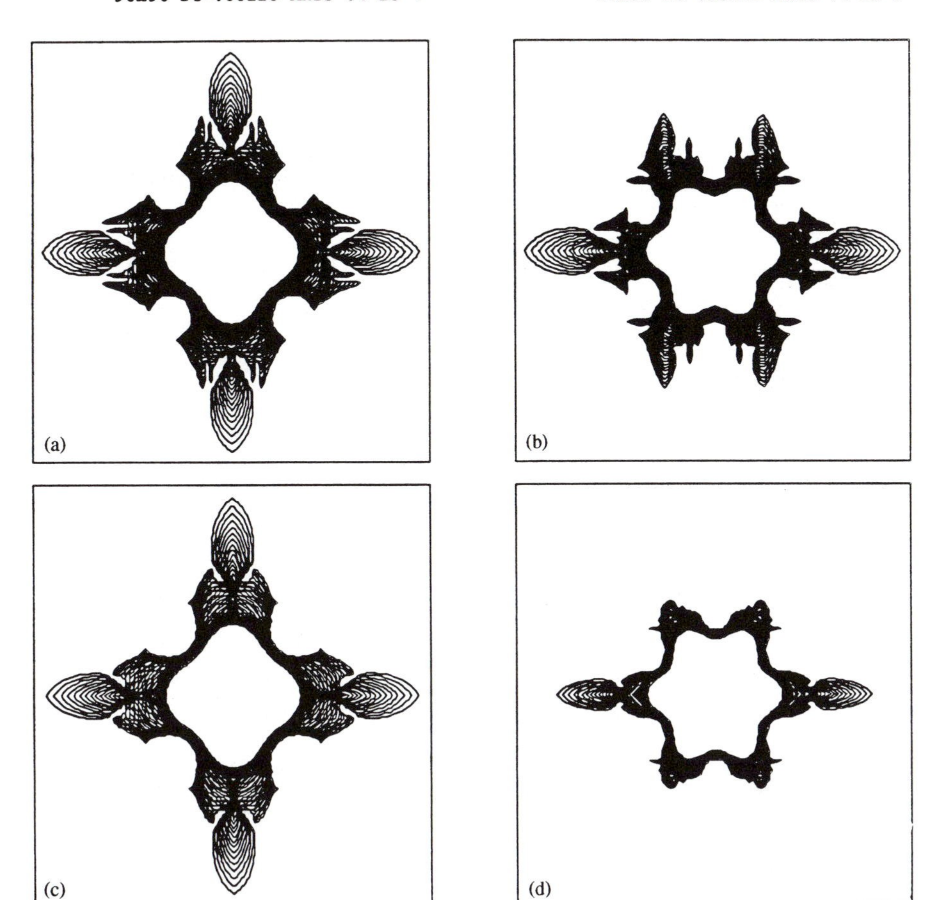

Figure 7: Changing Initial Perturbation, Anisotropy Coefficient A, and Crystalline Symmetry k_A

Fig. 7a: $L = 4, A = 0.4, k_A = 4, \varepsilon_C = 0.001, \varepsilon_V = 0.001, H = 1.0, 96 \times 96, \Delta t = 0.00125$

Fig. 7b: $L = 4, A = 0.4, k_A = 6, \varepsilon_C = 0.001, \varepsilon_V = 0.001, H = 1.0, 96 \times 96, \Delta t = 0.00125$

Fig. 7c: $L = 4, A = 0.8, k_A = 4, \varepsilon_C = 0.001, \varepsilon_V = 0.001, H = 1.0, 96 \times 96, \Delta t = 0.00125$

Fig. 7d: $L = 6, A = 08., k_A = 6, \varepsilon_C = 0.001, \varepsilon_V = 0.001, H = 1.0, 96 \times 96, \Delta t = 0.00125$

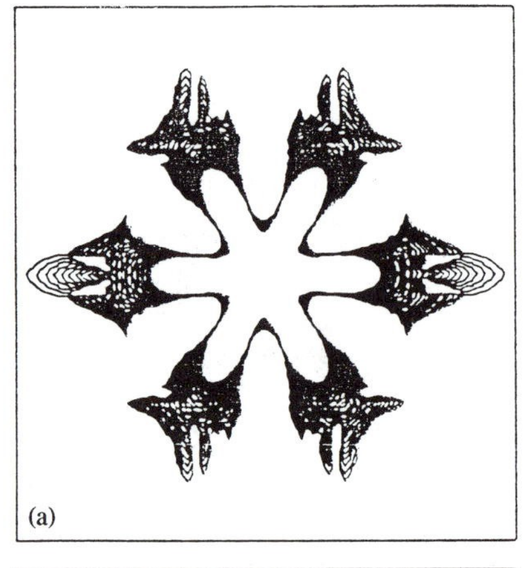

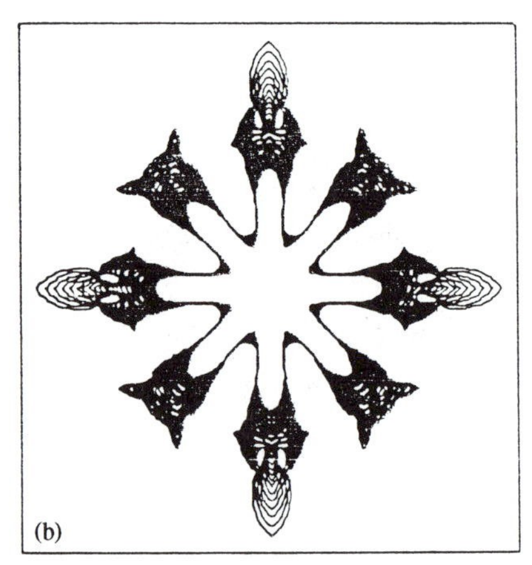

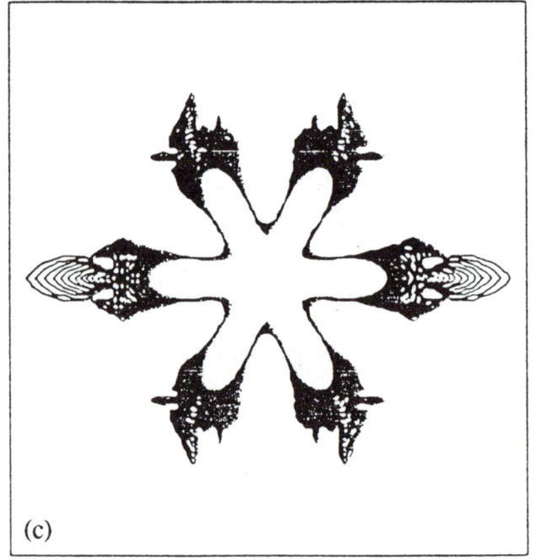

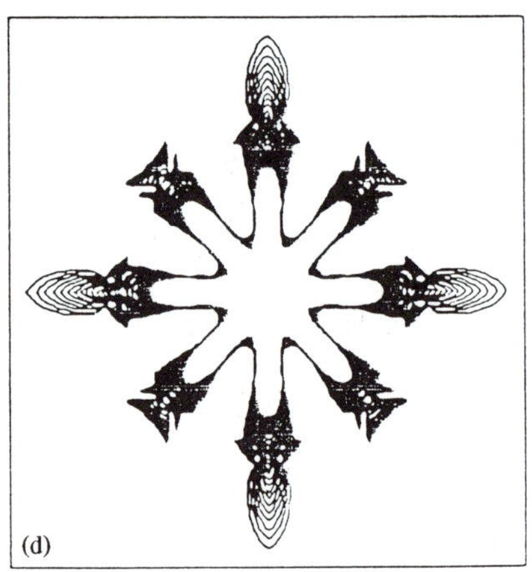

Figure 8: Changing Initial Perturbation, Anisotropy Coefficient A, and Crystalline Symmetry K_A

Fig. 8a: $L = 6, A = 0.0, k_A = 6, \varepsilon_C = 0.001, \varepsilon_V = 0.001, H = 1.0, 96 \times 96, \Delta t = 0.00125$

Fig. 8b: $L = 4, A = 0.4, k_A = 6, \varepsilon_C = 0.001, \varepsilon_V = 0.001, H = 1.0, 96 \times 96, \Delta t = 0.00125$

Fig. 8c: $L = 4, A = 0.8, k_A = 4, \varepsilon_C = 0.001, \varepsilon_V = 0.001, H = 1.0, 96 \times 96, \Delta t = 0.00125$

Fig. 8d: $L = 6, A = 0.8, k_A = 6, \varepsilon_C = 0.001, \varepsilon_V = 0.001, H = 1.0, 96 \times 96, \Delta t = 0.00125$

REFERENCES

[1] BRUSH, L.N., AND SEKERKA, R.F., *A Numerical Study of Two-Dimensional Crystal Growth Forms in the Presence of Anisotropic Growth Kinetics*, J. Crystal Growth, to appear.

[2] CAHN, J. W., AND HILLIARD, J. E., Jour. Chem. Phys. 28 (1958), pp. 258.

[3] CHORIN, A.J., *Curvature and Solidification*, Jour. Comp. Phys., 57 (1985), pp. 472–490.

[4] EVANS, L.C., AND SPRUCK, J., *Motion of Level Sets by Mean Curvature, I*, J. Diff. Geom., (to appear).

[5] GREENGARD, L., AND STRAIN, J,, *A Fast Algorithm for Heat Potentials*, Comm. Pure Appl. Math XLIII,(1990) 949–963.

[6] GURTIN, M.E., *On the Two-Phase Stefan Problem with Interfacial, Energy and Entropy*, Arch. Rat. Mech. Anal., 96 (1986), pp. 199–241,.

[7] KARMA, A., *Wavelength Selection in Directional Solidification*, Phys. Rev. Lett., 57 (1986), pp. 858–861.

[8] KELLY, F. X., AND UNGAR, L.H., *Steady and Oscillatory Cellular Morphologies in Rapid Solidification*, Phys. Rev. B, 34 (1986), pp. 1746–1753.

[9] KESSLER, D.A., AND LEVINE, H., *Stability of Dendritic Crystals*, Phys. Rev. Lett., 57 (1986), pp. 3069–3072.

[10] LANGER, J. S., *Instabilities and Pattern Formation in Crystal Growth*, Rev.Mod.Phys., 52, pp. 1–28 1980 Linear Stability, Crystal Growth,.

[11] MEIRON, D.I., *Boundary Integral Formulation Of The Two-Dimensional, Symmetric Model Of Dendritic Growth*, Physica D, 23 (1986), pp. 329–339,.

[12] MEYER, G. H., *Multidimensional Stefan problems*, SIAM J. Numer. Anal., 10 (1973), pp. 552–538,.

[13] MULDER, W., OSHER, S.J., AND SETHIAN, J.A., *Computing Interface Motion: The Rayleigh-Taylor and Kelvin-Helmholtz Instabilities*, to appear, Journal of Computational Physics, 1991.

[14] MULLINS, W. W., AND SEKERKA, R. F., *Morphological Stability of a Particle Growing by Diffusion or Heat Flow*, Jour. Appl. Phys., 34 (1963), pp. 323–329,.

[15] MULLINS, W. W., AND SEKERKA, R. F., *Stability of a Planar Interface During Solidification, of a Dilute Binary Alloy*, Jour. Appl. Phys., 35 (1964), pp. 444–451,.

[16] OSHER, S.J., AND SETHIAN, J.A., *Fronts Propagating with Curvature-Dependent Speed: Algorithms based on Hamilton-Jacobi Formulations*, J. Comp. Phys., 79 (1988), pp.12–49.

[17] SETHIAN, J.A., *Curvature and the Evolution of Fronts*, Communications of Mathematical Physics, 101 (1985), 4.

[18] SETHIAN, J.A., *A Review of Recent Numerical Algorithms for Hypersurfaces Moving with Curvature-Dependent Speed*, J. of Diff. Geom., 31 (1989), pp. 131–161.

[19] SETHIAN, J.A., AND CHOPP, D.L, *Singularity Formation and Minimal Surfaces*, (to appear).

[20] SETHIAN, J.A., AND STRAIN, J., *Crystal Growth and Dendritic Solidification*, to appear, J. Comp. Phys. (1991).

[21] SETHIAN, J.A., AND STRAIN, J., *Three-Dimensional Crystal Growth*, in preparation.

[22] SMITH, J. B., *Shape Instabilities and Pattern Formation in Solidification : A New Method for Numerical Solution of the Moving Boundary Problem*, Jour. Comp. Phys., 39 (1981), pp. 112–127.

[23] STRAIN, J., *A Boundary Integral Approach to Unstable Solidification*, J. Comp. Phys., 85 (1989), pp. 342-389.

[24] STRAIN, J., *Fast Potential Theory II: Layer Potentials and Discrete Sums*, to appear, J. Comp. Phys..

[25] STRAIN, J., *Linear Stability of Planar Solidification Fronts*, Physica D, 30 (1988), pp. 297–320.

[26] STRAIN, J., *Velocity Effects in Unstable Solidification*, SIAM Jour. Appl. Math. (1990).

[27] STRAIN, J., *A Fast Boundary Integral Method for Crystal Growth*, in preparation.

[28] SULLIVAN, J. M., LYNCH, D. R., AND O'NEILL, K. O., *Finite Element Simulation of Planar Instabilities, during Solidification of an Undercooled Melt*, Jour. Comp. Phys., 69 (1987), pp. 81–111.

TOWARDS A PHASE FIELD MODEL FOR PHASE TRANSITIONS IN BINARY ALLOYS

A. A. WHEELER* AND W. J. BOETTINGER†

Abstract. To date phase field models have only been used to model non-isothermal phase transitions in a pure material. Here we describe recent steps which aim to extend phase field models to deal with binary alloys; a situation of metallurgical and industrial importance. To this end we present a new phase field model for isothermal phase transitions in a binary alloy and discuss the results of an asymptotic analysis. Finally we suggest ways in which these models may be further developed to achieve our aim of a non-isothermal phase field model of a binary alloy.

1. Introduction

Classical macroscopic models of phase transitions model the interface between regions of different phase as a surface, and hence assume it has zero thickness. The governing equations for thermodynamic variables, such as temperature and solute, are formulated in each phase independently, based upon conservation principles and quantitatively verified phenomenological laws. The boundary conditions at the interface are chosen to describe the processes, such as liberation of latent heat and segregation that occur at the interface. This approach gives rise to the formulation of a free boundary problem which provides a difficult mathematical setting; only the simplest models of phase change have been rigorously mathematically analyzed. Because these models have been used for many years, it is clear from the outset what physical mechanisms are incorporated into them, and comparison with careful controlled experiments is possible.

An alternative technique for investigating transport processes in systems undergoing a phase transition involves the construction of a Landau-Ginzberg free energy functional. This approach has its roots in statistical physics (Landau and Khalatinikov, [1]). Further, a phase field, which is a function $\phi(\mathbf{x},t)$, is postulated which describes the phase of the system at any point in time and space. It is assumed that the Helmholtz free-energy $\mathcal{G}(\phi,\ldots)$, is a functional of the phase-field, as well as any other thermodynamic variables, (such at temperature which are denoted here by ellipsis) in the following way:

$$\mathcal{G}(\phi,\ldots)=\int_\Omega \left\{\tfrac{1}{2}\epsilon^2(\nabla\phi)^2+g(\phi,\ldots)\right\}d\Omega, \tag{1}$$

where Ω is the region occupied by the system, and $g(\phi,\ldots)$ is the free energy density. Its dependence on ϕ usually has a "double well" form. The phase field is then assumed to evolve as:

$$\dot{\phi}\propto L\left(\frac{\delta\mathcal{G}}{\delta\phi}\right), \tag{2}$$

where L is some partial differential operator. This equation is then supplemented by partial differential equations for the other thermodynamic variables. Cahn and Hilliard [2] have successfully used this approach to model spinodal decomposition of a binary alloy, although here the concentration naturally plays the role of the phase field. Various models that employ this idea are reviewed by Halperin, Hohenburg and Ma

* Permanent address: School of Mathematics, University Walk, Bristol, BS8 1TW, UK
† National Institute of Standards and Technology, Gaithersburg, MD 20899 USA

[4], particularly in regard to the study of critical phenomena. The Model C given by these authors has been adapted by Langer [5], Fix [6] and most prolifically by Caginalp [7], to derive the so-called "phase-field model" of solidification which models the phase change of a pure material. Caginalp has extensively studied this, and variations of this model [8,9]. It has emerged from study of this model that qualitatively it exhibits features common to solidification of a pure material. Numerical calculations based on this model, by Smith [10], and a similar model, by Kobayashi [11], show breakdown of a planar and circular interfaces to cellular structures, as well as the formation of dendrite-like structures, inclusion of liquid pockets, and coarsening behavior.

Caginalp [8] has shown in various distinguished limits, in which $\epsilon \to 0$, that various forms of the classical Stefan problem may be recovered, in which the interface is taken to be "sharp," i.e., modeled by a surface. In this limit there are thin layers within Ω of thickness $\mathcal{O}(\epsilon)$ in which the phase field rapidly changes. These are interpreted as representing interfaces, which are necessarily diffuse. From this analysis it transpires that in some limits, the interfacial dynamics involve curvature effects corresponding to the Gibbs-Thomson interfacial surface energy as well as kinetic effects. Further, it is also possible to recover the classical Hele-Shaw problem in other limits. It is clear that this approach can embody a considerable variety of realistic physical effects in a coherent way.

Difficulty with this particular model, as pointed out by Penrose and Fife [12], is that its derivation is thermodynamically inconsistent. This is because the free energy functional is *only* employed in the formulation of the kinetic equation for the phase field. The concern here is that the solution of the above governing equations does not correspond to the total free energy of the system decreasing monotonically with time, as required by the Second Law of Thermodynamics. An alternative approach suggested by these authors is to construct an entropy functional, $\mathcal{S}$, of the system and require it to evolve as a gradient flow of the form

$$\dot{\mathbf{u}} \propto \mathrm{grad}_0 \mathcal{S}(\mathbf{u}), \tag{3}$$

where grad_0 is a suitable constrained gradient, and $\mathbf{u}$ represents the thermodynamic variables. This formulation necessarily ensures that the total entropy of the system increases with time.

The appeal of phase field models in describing phase transitions is twofold;

- It provides a simple, elegant description, that appears to embody a rich variety of realistic physical phenomena.
- From a computational point of view it is relatively simple to compute solutions. This is because it is not necessary to distinguish between the different phases. Computations on the classical sharp interface formulation require that the free boundary is tracked numerically and that the region occupied by each phase is therefore determined and dealt with individually. This results in very difficult and untidy numerical algorithms.

Our aim is to extend the phase field models developed so far for pure materials to binary alloys. As a first step in achieving this goal we have developed a phase field model of phase transitions in an *isothermal* binary alloy. In section 2 we describe this new model and in section 3 we discuss the properties of this model, particularly in regard to the results of an asymptotic analysis of it conducted by Wheeler, Boettinger and McFadden [14], henceforth referred to as WBM. In section 4 we outline possible generalizations of our phase field model and suggest a tentative phase field model for a non-isothermal binary alloy.

2. Phase Field Model of an Isothermal Binary Alloy

We consider an isothermal solution of two different species A and B in which are present two phases, solid and liquid, contained in a fixed region Ω with boundary $\partial\Omega$. We denote the concentration of B by $c(\mathbf{x},t)$ and we introduce a phase field $\phi(\mathbf{x},t)$ which represents the phase in time and space in Ω. For definiteness we describe the solid-liquid interface by $\phi(\mathbf{x},t) = \frac{1}{2}$ and denote solid regions where $\phi(\mathbf{x},t) > \frac{1}{2}$ and liquid regions where $\phi(\mathbf{x},t) < \frac{1}{2}$.

A recent phase field model due to Kobayashi [11] models the phase transition of a pure material by employing the following gradient weighted free-energy functional:

$$\mathcal{F}_K(\phi,T) = \int_\Omega \left\{ \frac{\epsilon^2}{2}|\nabla\phi|^2 + f_K(\phi,T) \right\} d\Omega, \tag{4}$$

where ϵ is a constant, $T(\mathbf{x},t)$ is the temperature and the free energy density $f_K(\phi,T)$ is represented by

$$f_K(\phi,T) = \int^\phi p(p-1)(p-\frac{1}{2}-\beta(T))dp, \tag{5}$$

where $\beta(T)$ is a monotonic increasing function of T, such that $\beta(T_M)=0$, where T_M is the freezing temperature of the material and $|\beta(T)| < \frac{1}{2}$. The free energy density $f_K(\phi,T)$ is a double well potential. The restriction $|\beta(T)| < \frac{1}{2}$ ensures that it has local minima at $\phi=0$ and $\phi=1$, and a local maxima at $\phi = \frac{1}{2}+\beta(T)$. Because of the two minima the system may exist stably in a state which is all-liquid ($\phi(\mathbf{x},t)\equiv 0$) or all-solid ($\phi(\mathbf{x},t)\equiv 1$). There is an energy penalty for a change of phase within the region Ω, which corresponds to ϕ varying between zero and unity. This is because such a variation increases the total energy $\mathcal{F}_K$ of the system, due to an increased energy density associated with the double well nature of the energy density, and also due to the contribution to the total energy due to the gradient energy, which is no longer zero.

If $-\frac{1}{2} < \beta(T) < 0$, then the *global* minima of the energy density is at $\phi = 1$ and so the all-solid state is the one with the lowest energy and is hence the preferred state. However, if $0 < \beta(T) < \frac{1}{2}$ then the situation is reversed and the liquid is the preferred state. We see that at temperatures below the melting point the solid phase has the minimum energy and is preferred, whereas for temperatures above the melting temperature, the all-liquid phase is preferred.

We now employ this form for the free energy density to develop the appropriate free energy density for an isothermal ideal solution. We assume that the temperature, T, which is given, is such that if the solution consisted only of species A ($c\equiv 0$) the all-solid phase would be the preferred state, i.e., $T < T_M^A$, where T_M^A is the melting temperature of pure A. Further, we also assume that the temperature T is sufficiently large that if the system consisted of species B alone ($c \equiv 1$) the all-liquid phase would be the preferred state, i.e., $T > T_M^B$, where T_M^B is the melting temperature of pure B. We also assume that the *molar* Gibbs free energy densities of each species A and B alone are of the form given by Kobayashi, and are denoted by $f_A(\phi;T)$ and $f_B(\phi;T)$ respectively. Specifically we put:

$$f_A(\phi;T) = W_A \int^{\phi} p(p-1)(p-\frac{1}{2}-\beta_A(T))dp, \tag{6}$$

$$f_B(\phi;T) = W_B \int^{\phi} p(p-1)(p-\frac{1}{2}-\beta_B(T))dp, \tag{7}$$

where here W_A, W_B are constants, and T the temperature is a parameter in this isothermal situation. We note that because $T_M^B < T < T_M^A$, then $-\frac{1}{2} < \beta_A(T) < 0 < \beta_B(T) < \frac{1}{2}$. The total energy density $f(\phi, c; T)$ of the solution is:

$$f(\phi,c;T) = cf_B(\phi;T) + (1-c)f_A(\phi;T) + \frac{RT}{v_m}[c\ \log c + (1-c)\ \log(1-c)], \tag{8}$$

where R is the gas constant constant and v_m is the molar volume. The first two terms correspond to the contribution to the energy density due to the individual molar Gibbs free energies of the two species and the last term is due to the decrease in energy associated with the mixing of the two constituents, under our assumption that it is an ideal solution. The free-energy density is illustrated as a function of c and ϕ in Figure 1.

In a similar way to Kobayashi we define the free energy functional by:

$$\mathcal{F}(\phi,c;T) = \int_{\Omega} \left\{ \frac{\epsilon^2}{2}|\nabla\phi|^2 + f(\phi,c;T) \right\} d\Omega. \tag{9}$$

In order to derive a kinetic model we make the assumption that the system evolves in time so that its total free-energy decreases monotonically. This may be met by assuming the rate of change of c and ϕ vary according to the constrained gradient of $\mathcal{F}(\phi,c;T)$:

$$\dot{\mathbf{u}} \propto -\text{grad}_0 \mathcal{F}(\mathbf{u}), \tag{10}$$

where $\mathbf{u} = (\phi, c)^T$. Fife [13] discusses how such constrained gradients may defined in a more rigorous mathematical setting. The only constraint we require here is that composition is conserved, i.e., $d/dt \int_{\Omega} c d\Omega = 0$. We chose the constrained gradient such that:

$$\frac{\partial\phi}{\partial t} = -M_1 \frac{\delta\mathcal{F}}{\delta\phi}, \tag{11}$$

$$\frac{\partial c}{\partial t} = M_2 \nabla \cdot \left\{ c(1-c)\nabla\frac{\delta\mathcal{F}}{\delta c} \right\}, \tag{12}$$

where M_1 and M_2 are constants. The boundary conditions are

$$\frac{\partial\phi}{\partial n} = \frac{\partial c}{\partial n} = 0, \tag{13}$$

where $\mathbf{n}$ is the outward normal to the boundary $\partial\Omega$. We may interpret the right hand side of (12) as the negative of the divergence of solute flux, $\mathbf{j} = -M_2\, c(1-c)\nabla\delta\mathcal{F}/\delta c$. The coefficient $c(1-c)$ has been included to ensure that the diffusion equation for the solute that emerges has a diffusion coefficient that is constant.

Evaluating the variational derivatives of the free energy functional gives that

$$\frac{\partial\phi}{\partial t} = M_1 \left(\epsilon^2\nabla^2\phi - \frac{\partial f}{\partial\phi} \right), \tag{14}$$

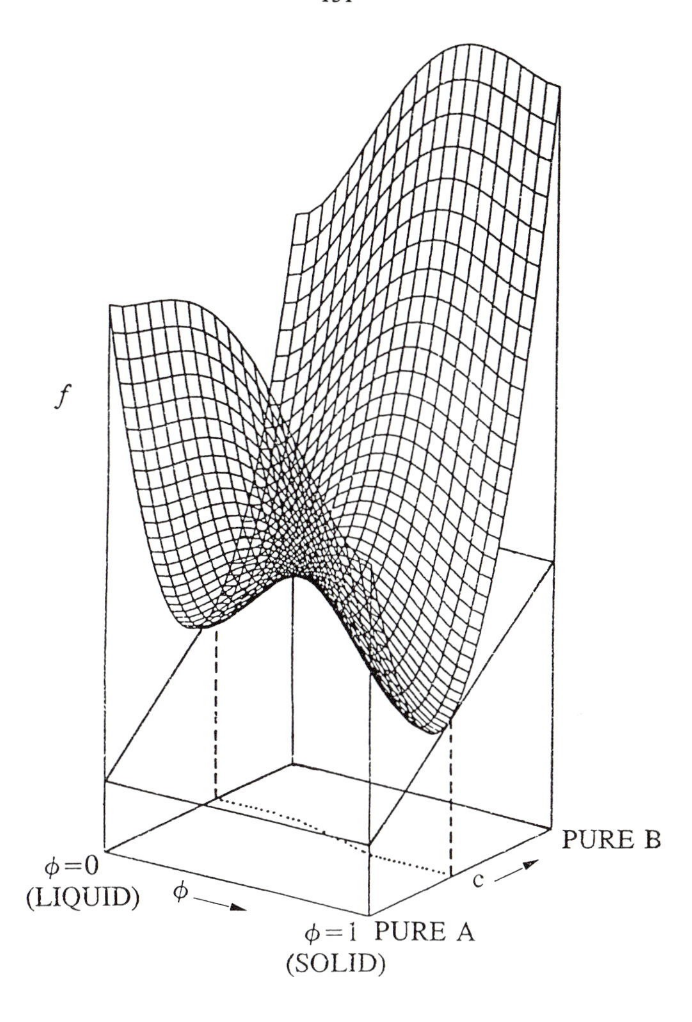

Figure 1: The free-energy density function $f(\phi, c)$, showing the common tangent plane construction which determines the interfacial concentration when the interface is stationary.

$$\frac{\partial c}{\partial t} = M_2 \nabla \cdot \left(c(1-c) \nabla \frac{\partial f}{\partial c} \right), \tag{15}$$

which may be also written as:

$$\frac{\partial \phi}{\partial t} = M_1 \left[\epsilon^2 \nabla^2 \phi - \left(c \frac{\partial f_A}{\partial \phi} + (1-c) \frac{\partial f_B}{\partial \phi} \right) \right], \tag{16}$$

$$\frac{\partial c}{\partial t} = M_2 \nabla \cdot (c(1-c) \nabla (f_A - f_B)) + D \nabla^2 c, \tag{17}$$

where $D = M_2 RT/v_m$ is the diffusivity of the solute (here taken to be constant).

A dimensionless form of the governing equations, based on the diffusive time scale is

$$\frac{\partial \phi}{\partial \tilde{t}} = \tilde{M}_1 \left(\tilde{\epsilon}^2 \tilde{\nabla}^2 \phi - \frac{\partial \tilde{f}}{\partial \phi} \right), \tag{18}$$

$$\frac{\partial c}{\partial \tilde{t}} = \tilde{\nabla} \cdot \left(c(1-c) \tilde{\nabla} \frac{\partial \tilde{f}}{\partial c} \right), \tag{19}$$

where

$$\tilde{M}_1 = \frac{M_1 l^{*2} RT^*}{D v_m}, \quad \tilde{\epsilon} = \frac{\epsilon}{l^*} \sqrt{\frac{v_m}{RT^*}}; \tag{20}$$

here l^* is a characteristic length of the interface. Henceforth for notational simplicity we will omit the tildes.

A particular difficulty in employing phase field models to real situations is relating their dimensionless control parameters, here ϵ, M_1, W_A, W_B and the functions $\beta_A(T)$ and $\beta_B(T)$ to real material parameters and growth conditions. WBM address this question and show that knowledge of the surface energy, solute diffusivity, interfacial thickness as well as the dependence of the free-energy difference on temperature provides a unique determination of the quantities $\epsilon, M_1, W_A, W_B, \beta_A(T)$ and $\beta_B(T)$. They illustrate this for the particular case of a Nickel-Copper alloy.

3. Asymptotic Analysis

WBM conducted an asymptotic analysis of the governing equations (18) and (19) in the limit $\epsilon \to 0$ with $M_1 = M/\epsilon$. From their consideration of the control parameters for the Nickel-Copper alloy this is a physically realistic limit.

They show that the system evolves on three different time scales; an initial rapid transient $\mathcal{O}(\epsilon)$, an order one time scale, and a final slow time scale $\mathcal{O}(\epsilon^{-1})$. We now summarize the behavior on each time scale in turn.

3.1. Interfacial genesis: $t = \mathcal{O}(\epsilon)$. Assuming that the initial data for both ϕ and c lies in $[0,1]$ then during this fast initial phase the system evolves to lower its energy by the phase field evolving to zero or unity in a point-wise manner, depending on whether the initial value of the phase field at a particular point in space lies in the region of attraction of zero or one. The evolution of the phase field is controlled by the ordinary differential equation:

$$\frac{\partial \phi}{\partial t} = -M \frac{\partial f}{\partial \phi}$$

with

$$\phi = \phi_0(\mathbf{x}) \text{ at } t = 0,$$

where $\phi_0(\mathbf{x})$ is the initial data. The concentration is essentially unchanged during this period and solid and liquid regions become differentiated. It therefore corresponds to the birth of the interfacial layers and is a consequence of the assumption that the mobility of the phase field is much larger than the diffusivity of the solute, in which case solute diffusion is unable to respond quickly enough to the changes in phase. The system acts to lower its free-energy solely by generating regions of solid or liquid.

3.2. Diffusive regime: $t = \mathcal{O}(1)$. During this period, the interfacial layers born during the initial rapid transient move in response to changes in the solute concentrations in the liquid and solid phase that evolve under the influence of solute diffusion. In fact the sharp interface model that is recovered from the asymptotic analysis for the leading order solute and concentration field $c^{(0)}$ during this period is:

$$\frac{\partial c^{(0)}}{\partial t} = \nabla^2 c^{(0)}, \quad \phi^{(0)} = 0 \text{ or } 1,$$

with boundary conditions at the interfacial surfaces

$$\left. \frac{\partial c^{(0)}}{\partial n} \right|_{liquid}^{solid} = -v_n c^{(0)} \Big|_{liquid}^{solid} \tag{21}$$

and

$$c^{(0)}_{solid} = c^{(0)}_S(v_n) \text{ and } c^{(0)}_{liquid} = c^{(0)}_L(v_n). \tag{22}$$

Here $c^{(0)}_{solid}$ and $c^{(0)}_{liquid}$ are the leading order interfacial solute concentrations in the solid and liquid phase respectively and v_n is the normal interfacial velocity assumed positive when the interface advances into the liquid. The boundary condition (21) is simply a statement of conservation of solute across the interface and is a consequence of the conservative form of the governing equation for the solute.

The functions $c^{(0)}_S(v_n)$ and $c^{(0)}_L(v_n)$, which represent the dependence of the interfacial concentrations on the normal interface velocity, have to be computed numerically and are related to the detailed structure of the phase field in the interfacial layers. It can be shown that this dependence is consistent with the interfacial layers moving so that they reduce the energy of the system by converting regions of high energy density to a lower energy density.

Moreover the forms of the functions $c^{(0)}_S(v_n)$ and $c^{(0)}_L(v_n)$ are fully equivalent to the following geometric construction: When the interface is stationary $c^{(0)}_{solid}$ and $c^{(0)}_{liquid}$ are determined by the values of $c^{(0)}$ common to the surface $z = f(c^{(0)}, \phi^{(0)})$ and the tangent plane to the surface $z = f(c^{(0)}, \phi^{(0)})$ which is tangent to the surface at *both* $\phi = 0$ and $\phi = 1$. This geometric construction is also shown in Figure 1. It may be interpreted in thermodynamic terms as requiring the chemical potential to be continuous across the interface. When the interface is not stationary this geometric construction is modified by requiring that the appropriate concentrations are given by two parallel tangent planes which are separated by a distance that depends on the value of v_n. When $v_n = 0$ this distance is zero and this construction reduces to the one given above.

This period of evolution is thus characterized by interfacial motion in response to energy differences across them and solute redistribution in the solid and liquid regions due to diffusion. The motion of the interfaces decreases with time and the energy of the system approaches it final minimum value.

3.3. Coarsening regime: $t = \mathcal{O}(\epsilon^{-1})$. When the system has energy close to its final value at the end of the previous period the energy differences driving the interfacial motion are small and so the system evolves very slowly. During this period the normal interface velocity is $\mathcal{O}(\epsilon)$ and the interfacial motion is controlled by the first order concentration field which is also $\mathcal{O}(\epsilon)$, as the concentration in each phase is constant to leading order. It is during this period that the interfacial curvature is important. The first order interfacial concentration in the liquid is given by:

$$c^{(1)}_{liquid}\left(\frac{c^{(0)}_{liquid}-c^{(0)}_{solid}}{c^{(0)}_{solid}(1-c^{(0)}_{liquid})}\right)+\Gamma_0(\mathcal{K}+M^{-1}\bar{V}_n)=0, \tag{23}$$

where $\bar{V}_n$ is the normal interfacial velocity on the slow time scale.

During this period the interfaces move in response not only to the small $\mathcal{O}(\epsilon)$ energy differences across them, i.e., thermodynamic disequilibrium, but also to the surface energy associated with curvature, which is also $\mathcal{O}(\epsilon)$. This balance of the thermodynamic equilibrium against surface energy indicates that this final long time evolution represents Ostwald ripening, in which a slow coarsening of the system is to be expected.

WBM also conducted numerical integrations of the governing equations that qualitatively confirm the results of the asymptotic analysis.

4. Discussion

The phase field model for the binary alloy given above was devised to provide the simplest description of an isothermal phase transition of a binary alloy. It therefore restricted its attention to the unrealistic situation where the solute diffusivity is the same in both solid and liquid phases. This restriction is simply overcome by allowing M_2 to be a linear function of ϕ.

We also limited our model to an ideal solution. It can be easily extended to the case of a regular solution by adding a term of the form $c(1-c)[\Omega_L + \phi(\Omega_S - \Omega_L)]$ to the free-energy density $f(c,\phi)$. It should then be a straightforward matter to modify the asymptotic analysis given above.

The above modifications are straightforward and are attempts to make our model applicable to a wider, more interesting class of materials. A more serious defect in the model, as it presently stands, is its prediction of the dependence of the segregation coefficient, $k = c_{solid}/c_{liquid}$, on interface velocity. WBM show that the model has the property that the segregation coefficient decreases as the interface velocity v_n increases. It is known experimentally that the opposite is true, and that in particular at high growth rates the segregation coefficient increases, and approaches unity. The behavior given by our model is a direct consequence of the geometric parallel tangent plane construction for the determination of the interfacial concentrations. In our model no account is taken of contributions to the total energy due to the concentration gradients. When the interface motion is rapid large concentration gradients will be established and they will contribute significantly to the energy of the system. Thus we anticipate that inclusion of a solute gradient energy is required to properly treat the velocity dependence of the segregation coefficient. This is the subject of current research.

Our aim is to develop a phase field model for phase transitions in a non-isothermal binary alloy. There now exist two separate phase field models that deal with aspects of this situation: The phase field model for a pure material models the heat transport (albeit for zero concentration) and our new phase field model discussed above provides a description of the solute transport. In order to develop a phase field model for a binary alloy these two models need to be combined. The simplest way in which to do this is just to append to the governing equations of our model (11) and (12), the heat equation used by the phase field model for a pure material:

$$\frac{\partial T}{\partial t} = \kappa \nabla^2 T + l\phi_t,$$

where κ is the thermal diffusivity and $l = L/(\rho c_p)$, L is the latent heat per unit mass, ρ is the density and c_p is the specific heat. Without a detailed investigation it is not clear that such an approach will be successful.

A more satisfactory approach, that also recognizes the thermodynamic difficulties associated with the phase field model of a pure material discussed by Penrose and Fife [12], would be to use the Gibbs free-energy density employed in our model and the entropy density employed by Penrose and Fife to discuss the phase field model of a pure material, to construct a new entropy density appropriate to a non-isothermal binary alloy system. With this in hand a thermodynamically consistent model could then be derived which ensures that the entropy increases monotonically with time. The resulting governing equations could then be approximated in a rational way in order to provide a simplified model which, for instance, recovered the linear heat equation in each phase.

REFERENCES

[1] L. D. Landau and Khalatnikov, I. M., in; Collected works of L. D. Landau, ed. D. ter Haar (Pergamon, Oxford, 1965) pp. 626 -633.

[2] J. W. Cahn and Hilliard, J. E., J. Chem. Phys. **28** (1958) 258-267.

[3] J. W. Cahn, Acta Metallugica, **9** (1961) 795-801.

[4] B. I. Halperin, Hohenburg, P.C. and Ma, S.-K., Phys. Rev. B. **10** (1974) 139-153.

[5] J. S. Langer, in: Directions in Condensed Matter Physics, (World Science Publishers, 1986) pp. 164-186.

[6] G. Fix, in: Free Boundary Problems, ed. A. Fasano and M. Primocerio, (Pitman, London, 1983) pp. 580-589.

[7] G. Caginalp, Arch. Rat. Mech. Anal. **92** (1986) 205-245.

[8] G. Caginalp, Phys. Rev. A. **39** (1989) 5887-5896.

[9] G. Caginalp and Fife, P. C., Phys. Rev. B. **33** (1986) 7792-7794.

[10] J. Smith, J. Comp. Phys. **39** (1981) 112-127.

[11] R. Kobayashi. Private communication (1990).

[12] O. Penrose and Fife, P. C., Physica D **43** (1990), 44-62.

[13] P. C. Fife, Proc. Taniguchi Int. Symp. on Nonlinear PDEs and Applications (Kinokuniya Pub. Co., 1990).

[14] A. A. Wheeler, W. J. Boettinger and G. B. McFadden, submitted to Phys. Rev. A.

JUL 28 1993